Practice Management for Land, Construction and Property Professionals

Practice Management for Land, Construction and Property Professionals

Edited by
Brian Greenhalgh
Head of External Affairs
School of the Built Environment
Liverpool John Moores University

Spon Press
an imprint of Taylor & Francis
LONDON AND NEW YORK

Preface

This book was originally conceived by members of the CPD (Continuing Professional Development) committee of the Royal Institution of Chartered Surveyors, after considering how we could increase the awareness of the necessity of developing business and managerial skills among the members. Early in 1995, it was decided to organize a conference on the theme of 'Practice Management', which would be appropriate not only to surveyors, but also to other members of the land, construction and property professions. Support was sought, and willingly given by the Royal Institute of British Architects, the Institution of Civil Engineers and the Chartered Institute of Building, who also all recognize the need for their members to develop more rounded skills in business management to cope with the ever-changing world of construction and property, where flexibility and responsiveness is as important as technical competence. E. & F.N. Spon also agreed to publish the material in book form, not as proceedings from the conference, but as edited material, and this volume is the consequence.

The conferences duly took place in London, on 28th September 1995, and in Liverpool on 5th October 1995, and the chapters in this book are the edited papers from these conferences. The reader is referred to the detailed list of papers and their authors which follows this Preface; in addition, my sincere thanks go to all who contributed their thoughts and research which have made this book the coherent package that it is.

The majority of the papers are written by members of the professions employed in academic institutions, and I also belong to this category. I make no apologies for this, since it is the function of academic institutions to research into modern practices, and present their findings to the profession at large, and for the practitioners and the institutions to consider these findings and react accordingly. What I find most gratifying is the evidence of collaboration between commercial organizations and universities in areas of professional practice management, which displays a leap in philosophy in how professional firms need to be 'managed' in the new competitive environment.

I am indebted to three people for the production of this book. Firstly, Joanna Edge of the RICS for her constant encouragement and help, Tom

Putt from the University of Reading, for reviewing and commenting on the papers, and finally to Paula Conboy from the JMU for her organizational and pestering abilities.

Brian Greenhalgh
Liverpool, December 1995

Papers presented at the conferences on 'Practice Management for Land, Construction and Property Professionals'

CONFERENCE 1: THE INSTITUTION OF CIVIL ENGINEERS, LONDON, 28TH SEPTEMBER 1995

The changing nature of professional work

Chairman: *Roy Swanston FRICS*
Keynote: *Michael Jeffries*
Changes in the construction industry, internally and externally generated
John Ridall, Sheffield Hallam University
Core values of construction professionals
David Root, University of Bath
A new management theory for modern professional firms
Tim Eccles, Kingston University
IT strategy and skills for the property professions
Dr Tim Dixon, College of Estate Management
The Chartered Surveyor as a management consultant
Keith Jones, DTZ Debenham Thorpe and Virginia Gibson, University of Reading

Managing professionalism and creativity

Chairman: *Professor Douglas Clelland*
Keynote: *Sir Michael Latham*
Creativity as a core skill
Mike Sharkey, University of Luton
The changing role of Human Resource Management
Dr Tom Kennie

Does structured training add value to professional organizations?
Simon Murray, University of Northumbria
Aims and benefits of being an Investor in People
Chris Hobbs, Surveyors Quality Management Services
Professional competence and occupational standards, their use in defining, assessing and certifying professional performance
Richard Larcombe, CISC
Team management applied to construction professionals
Andrew Charlett, Nottingham Trent University

CONFERENCE 2: MERSEYSIDE MARITIME MUSEUM, ALBERT DOCK, LIVERPOOL, 5TH OCTOBER 1995

Strategic management and marketing of professional services

Chairman: *Clive Lewis FRICS*
Keynote: *Professor Peter Lansley*
Creating a sustainable competitive advantage for property and construction companies
Professor Brian Sloan, Napier University, Edinburgh
The management of direct selling and pre-qualification team presentations for competitive advantage in contractual services
Dr Christopher Preece et al., *University of Leeds*
Changes to public sector procurement of property advice; the implications for private sector firms
Gaye Pottinger, College of Estate Management
The market for valuation services: applying the 'service/market matrix' to professional property services
Dr Peter Hobbs, Boots Properties Plc and Gavan O'Leary, University of the West of England
The client referral systems of UK building surveying practices
Michael Hoxley, University of Salford
A review of liability for construction and property firms
Stephen Donohoe, Singapore University.

Maintaining quality and professional ethics

Chairman: *Robin Wilson FEng FICE*
Keynote: *Ken Innes CEng FICE*
Professional ethics in construction
Lewis Anderson, University of Greenwich
Professional ethics, good business or sentimental liberalism?
Joseph Boughey and Chris Tremear, Liverpool John Moores University

PART ONE

The Changing Nature of Professional Work

Introduction to Part One

Michael Jeffries, Chief Executive, W S Atkins Ltd

By way of introduction to the remainder of this section, it is my intention to set the scene by looking briefly at the changes which have taken place in recent decades in the construction and property industries, explaining how for example, my firm W S Atkins has successfully analysed and responded to change over the last several years and how we see our business developing into the immediate future, and further, to look at one or two examples.

CHANGE IN GENERAL

What is change? We need to:

- recognize it
- understand it
- anticipate it
- manage it

It starts at a macro or international level; change which is world wide in its scope and impact. The economies of the principal trading blocs have been twice decimated in this century by world wars, and world peace for 50 years has enabled extraordinary economic progress to be made. What effects will the GATT agreement have? Sceptics believe that the improvements in world GDP by 2002 of $275 bn look impressive in absolute terms but spread over several years they represent less than 0.1% of the total – well within the level of estimating error.

At the turn of the century most of the world's top companies were founded on natural resources and the production of raw materials. Today, they are dominated by new technologies – pharmaceuticals, micro-electronics, telecommunications — where will they be tomorrow? The emergence of the huge global conglomerates has changed the balance of power. Thirty of the world's top companies now account for approximately 80% of the world's GNP.

Change then comes down to national level, which we have some prospect of understanding and measuring the impact upon our daily personal and business lives. Taxation, employment legislation, incentives, penalties, etc. all have a direct impact causing us to adjust what we do and how we do it.

Change also arises from technological development which in turn, of course, affects not only what we do and how we do it, but also alters the whole structure of industry. This, in my view, cannot be wholly separated from social change. In the United States most of the income gains have gone to the top 20% of the working population and 64% of these gains went to the top 1% in the 1980s. Currently, the bottom 60% has seen its income fall by around 20% in real terms. This pattern appears to be repeating itself in the UK and Europe and women have stepped into the breach. If they are married to men in the bottom 60% they have had to increase their hours to, in many cases, full time, and in future there will be no extra work effort available to offset a further reduction in male earnings.

This is all interesting background but I would like to get to a level to which we can all relate and look at recent changes which have directly affected the professions.

When I refer to my firm of W S Atkins being successful, I mean that since 1988 – when the recession first began to loom large – our firm has expanded its turnover to record levels in each of the subsequent years throughout the recession and, with the exception of 1991/92, has turned in record profits. Our turnover has increased from £59 m in 1987/88 to almost £200 m in 1994/95 and our staff numbers have risen from 1800 to over 5000 now, with much of the growth occurring in the last 3 years. However, what we do, and how we do it has changed significantly.

During my time in the profession, I have experienced the 'bear' period of the late 1960s followed by the boom in the early 1970s and then another downturn in the late 1970s and early 1980s. However, these periods were simply oscillations of the economic pendulum which have always occurred and the basic structure of life within the professions remained relatively unchanged. The period of the last 10 years, however, has without doubt been the most turbulent in terms of technological development and political change which in turn has had a significant effect upon the professions.

It is a matter of conjecture whether the Thatcher government created the climate for change in this country or whether it was simply one of the first administrations to recognize and respond to what was happening both nationally and internationally. Whether you agree or disagree with the policies, and how effect was given to them, there was unquestionably a need for radical reform and, by definition, this cannot be achieved by nibbling round the edges. It requires bold actions involving wholesale change and this is just what the government set about doing.

CHANGE IN THE PROFESSIONS

There were a number of drivers of change, with complex relationships between them, but perhaps those which have had the most profound effects upon the professions were in the areas of deregulation of the traditional professions, privatization and the introduction of competition between the public and private sectors. Britain had become internationally uncompetitive, and internally grossly inefficient; the level of spending necessary to continue to maintain and develop public infrastructure had become unsustainable against the competing demands for money from other areas. It was perceived that the construction industry was not performing as well as it might and the professions within it were not delivering value for money, particularly to the public sector.

Deregulation

The watershed was of course the deregulation of the construction industry professions in 1984 amid considerable protest and prophecies of disaster. This was done against a background at the time of increasing economic confidence which eventually materialized into the huge boom of the middle and late 1980s. There was much talk in the architectural profession about the quality of buildings which would emerge from this draconian regime. Ironically, after probably one of the worst periods in architectural history in the post-war years, the last 10 years in my view has seen some considerable improvement.

How did this competition affect W S Atkins? The firm had built up its business mainly in the public sector and had enjoyed patronage and scale fees like everyone else. At first we were extremely apprehensive, but we quickly realized that we were better equipped than most practices to cope with this significant change. Very simply, we already had a sophisticated (by comparison with others) management accounting system which enabled us to understand what our projects were costing us, including the attribution of all fixed and variable overheads. Even today, 11 years on, it is amazing how many quite large professional practices are still managed more or less on a cash basis. I know of two firms of several hundred staff, each of which have just got round to appointing a finance director in the last 12 months, and neither had much idea of their level of profitability except at each year-end.

Clearly, it is not possible to operate effectively in a highly competitive market if there is no understanding of the cost base and, of course, many firms, contractors and consultants, which were poorly managed have not survived the recession.

Deregulation and the public sector

As well as having a devastating effect upon the private sector, the backlash is now beginning to be felt in the public sector too. I doubt if anyone would disagree in principle with the notion that the public sector should seek to spend tax-payers' money in an efficient and cost-effective way. When the public sector first began to build up its own in-house professional and technical resources it was against a background of recognized professional fee scales and it was argued that the cost of these internal departments was less than employing external firms. However, following the deregulation of the professions, the market and the recession have driven down prices to almost unsustainable levels. Private firms have had to adjust to this situation and, as noted above, those that could not or did not, have failed.

Public/private sector competition

The public sector on the other hand has continued more or less as it was over the last several years. There have been some compulsory redundancies, but compared with the private sector they have been relatively slight. The result is a complete reversal of the situation that used to exist, i.e. it is now much less costly in absolute terms to commission the work in the private sector. The Government's response to this has been to introduce market testing in Central Government and seek to bring in legislation requiring Local Government to compulsorily tender their professional and technical services.

This looks to be a perfectly logical process on the face of it, but in fact there are many fundamental difficulties. Firstly, introducing competition between the public and private sectors is basically an unfair fight; the playing field can never be level and each side sees it tilted in favour of the other party. Secondly, the public sector is not equipped to deal with the effects of losing its 'market share'; it is restricted in other markets and cannot as easily reduce costs. Thirdly, it has to be recognized that the public sector has a great deal of specialist expertise, much of which relates to a detailed knowledge of the employing Authority's asset base; it is likely to have been built up over many years and could be lost by attrition.

Many Local Authorities have recognized the probable effects of Compulsory Competitive Tendering (CCT) and have sought to protect their staff and secure their ongoing involvement in future work (a good example of the public sector managing change). In fact, it could be argued that the Government's objectives are more likely to be achieved if this voluntary externalization continues. If and when CCT is introduced I suspect that, in overall terms, although work will be lost in competition to the private sector, there is likely to be a net increase in public sector costs arising out of a huge increase in the administration of the tendering process, exacerbated further by EU rules.

It must be recognized that CCT 'avoidance' is not simply the bidding of a bundle of existing and prospective contracts, it is the transfer of a business undertaking providing a whole raft of other services to the employer as well as those of a conventional professional and technical nature. It involves the transfer of public sector employees and their past service entitlements under the TUPE [Transfer of Undertakings (Protection of Employment)] Regulations of 1981 and recent court judgements make this area of the law a very uncertain and unsatisfactory basis for the assessment of contingent liabilities. It is fair to say that W S Atkins has learnt a great deal that is new in each one of the contracts it has concluded and the Board has recently instituted a complete reappraisal of the longer-term risks entered into by the firm to ensure that they are properly managed into the future.

W S Atkins' approach to the prospect of CCT has been to attack it from the outset. When the draft white papers were published in the late 1980s, we felt that the introduction of the legislation would be hugely damaging to the firm. Not that we feared competition from our peers; instead we believed that the Local Authorities would use all means at their disposal to hang on to their own work, tender for small parcels and that we would be priced out of the market. We quickly realized that many Authorities were thinking along similar lines and we initially became involved – along with the major management consultants — in advising Authorities and helping them to develop their thinking as to how to transfer their in-house resources to the private sector in a way which would exempt them from complying with the legislation, at least in the short term. Via this route we completed our first externalization at the end of 1989.

This approach has had a profound effect upon the small to medium-sized firms. They do not have the financial strength to take on the often considerable past services liabilities of public sector staff and as the larger firms have entered into 4- to 5-year contracts for all of the Authorities' work in exchange for accepting these liabilities, they have been effectively excluded from the market at least for that period and possibly for the foreseeable future.

Privatization

I have mentioned privatization as having had significant effects upon the professions and this has certainly been the case with W S Atkins. Our business had been built up in the public sector, including British Steel, National Coal Board and CEGB – all of which are now privatized and our work load with them is now negligible. However, as doors closed, others opened and we saw opportunities in the privatized water industry and in British Rail which otherwise would not have existed.

This leads on to another fundamental point. As the structure of the industry has changed so has the nature of the services it requires and the way in which it procures them. Quantity surveyors for example cannot survive on the Bills of Quantities type work which was their bread and butter 10 years and more ago. New skills have had to be developed and existing ones 'repackaged' and again, those firms which have failed to recognize and adapt to these changes have not survived.

Market factors

I have referred to the pricing of professional services at unsustainable levels. Although not wishing to complain, I am now reluctant to attend the annual dinners of the professional institutions because I am tired of hearing pleas to government ministers and the like to restore the status quo. It will never return. However, the market for professional services is far from being a settled one. It is still often impossible to properly and adequately define work scopes and we often find ourselves in competitions where the fee spread is several hundred per cent. This is, to say the least, most unsatisfactory but, if you believe that market forces will eventually regulate the market, there will come a time when this will become a thing of the past.

Businesses have always managed change. What is most problematical now is the hugely increased rate at which change takes place and the level of technical and commercial complexity which attaches to it.

The businesses which are going to survive are those which are able to analyse, understand and respond to the pressures of change, innovate and create change themselves, rather than those who resist it and hope that it will go away.

THE FUTURE

What of the future and how will it impinge upon the professions? Looking at our business, we see a number of things happening over the next few years which we will have to respond to.

1. The whole or partial funding of public sector infrastructure by the private sector. Whether or not the PFI survives in its present form, monies raised from taxation cannot cope with the funding demand and new ways will have to be found to solve this problem. Inevitably this will lead to long-term commercial relationships between the private and the public sectors which will transcend the colour of Governments. There is a huge misunderstanding in Government

about the transference of risk but hopefully this will gradually be resolved as more projects are concluded.

2. Technological change is continuous; information technology continues to develop and much of what we have is in its infancy. Communications already make it cost-effective to have detailed design work done remotely in other countries where the costs are a fraction of those here in Britain. Clients want high-level expertise and quality – this is the only basis upon which a premium on market rates can be achieved. Investment in technical and product development is absolutely essential, together with the management skills necessary to effectively exploit it.
3. Consultants need to add value to their clients' businesses. Clients increasingly are asking professional firms to share some of the risk with them and our problem is to manage the downside and find a means of participating in the reward.
4. Procurement and contracting is another area which is ripe for change. We already have client disenchantment with the conventional contractual relationships between the industry and the professions which has resulted in the huge increase in design and build. However, contractors today have few, if any, directly employed operatives and they make their money accepting risk, managing and organising sub-contract trades. We have already begun to look at this area and I believe there are certain parts of the market in which contractors will have difficulty in competing.
5. Managing people during these periods is vitally important. Change and uncertainty is demoralising and demotivating to some, just as it is exciting and challenging to others. Understanding motivation is difficult for those enthused as I am with the prospects ahead of us, knowing that we have built up the means to exploit them.

CHAPTER 1

The core values of construction professionals

1.1 INTRODUCTION

The first question, if we are to address professional ethics and professional values for construction and property professions must be to ask: 'What exactly are "professions"?'

The realistic answer is that we do not know for sure, we only really know what they are not – they are not the same as other occupations and they do not seem to have a great deal in common. The term 'a profession' is seen as a lay term with no precise denotation, there is no obvious explanatory trait linking together the various construction professions beyond the simple fact that they are called professions and that the term is imposed by society at large to identify a commonality which may not exist. Indeed, we identify closer with the terms 'architect', 'engineer' or 'surveyor' than any generic label of 'construction professional'.

But, are the construction professions different from other occupations? At the most basic level, the primary purpose of any professional, indeed of any practice, is to make money. In this they are no different from any other occupation and this fact ultimately underpins all behaviour. The difference to other occupations appears in the means by which this is done. The core professional values of honesty, integrity, etc. have merely evolved as the most effective means of doing this. Indeed, they are a rational response to the position practitioners may find themselves in, although the practitioner and the client are not aware of this as a 'rational decision' – that is, one of deliberate policy. Instead, it is a characteristic of the professional's identity and values. John Kay of the *Daily Telegraph* expressed it well:

> 'the doctor whose motivation is to make as much money as possible will not be a good doctor, and may well make less money than an able doctor whose motivation is the welfare of his patients'.

Studies of professions to date have looked at the application of a core body of knowledge as fundamental to the identity of profession. For

Carr-Saunders & Wilson, the precursor to the professional is the technical expert. For them the profession is a collection of technical experts with formal association and can be seen as an inevitable result of a historical process: the meeting of like-minded people in social situations, the discussion of common problems, attempts to resolve the problems ending with the formalization of these attempts and discussions into an organizational framework. Indeed a brief look at the history of professional associations such as the RICS, ICE or RIBA supports this view.

Normally we would consider this behaviour as a precursor to forming cartels, as Adam Smith observed in *The Wealth of Nations*;

> 'People of the same trade seldom meet together, even for merriment and diversion, but the conversation ends in a conspiracy against the public, or in some contrivance to raise prices'.

Yet with professions the situation seems to be different. Although some argue that the interests of the professions is that of the everyday concerns of interest politics, a desire for prestige or competition for control of bodies of knowledge with other occupations, all professions are formed with high-minded aims, to maintain standards of quality and service. Is this a cynical ploy? It seems likely that both sides benefit from the professions' existence: the practitioner or practice gain status, a 'selling point' to attach to their expertise and an institution to protect their interests (although they may not consciously recognize this), while the client gains protection through the control exercised by the professional associations and the professional values inherent in the associations' members.

1.2 THE CLIENT'S NEED FOR PROFESSIONAL VALUES

In professional practice (the primary sphere of occupation for professionals in the construction industry), the professional consultant is independent of the client and will only come into contact with him when the service is required. Thus the client is (generally) not in a position, unlike an employer, to judge the 'human aspects' of the service such as honesty or integrity.

For a client then, often relying on a total stranger, the worry is over the human aspects of the service. His question is one of ethics over policy; policy being a conscious choice between various possibilities, while ethics are derived from core attitudes or values. In other words, behaviour as an expression of unconscious values rather than optimization of outcomes.

This has been expressed by contrasting the person who is honest with the person who behaves honestly because it is the best policy. The first gives one more confidence in dealing with them, because their behaviour is predictable in the short/medium term (people do change their values, but not suddenly), while the second is responding to circumstance and their actions are predictable only if one knows what the factors are that affect them. Policy will change to suit the circumstances, and behaviour will therefore be unpredictable in changing circumstances.

The client needs to know that their interests will be protected. Legal remedies only relate to skill, not the personal traits of professionals, and cannot restrict a right to practice. The exclusion of an individual from the ranks of the 'privileged' members of the institution exists as a deterrent but can only be applied once the damage is done. It would be more effective to ensure that personal traits are suitable before entry. This is possibly the true role of the professional association: to control entry and ensure similar values through similar experiences of entry. Knowledge and values are developed from experience and by seeking to control the entry experience the professional associations control the acquisition of knowledge and values.

1.3 KNOWLEDGE AND VALUES

All construction, land and property professions are identified, but not determined, by two primary characteristics: (i) a prolonged period of training/education to acquire a specific body of knowledge; and (ii) a methodology to apply this knowledge to the ordinary business of life.

It has been accepted that knowledge by itself is insufficient and that the method of application of that knowledge (i.e. not only the technical but also moral and qualitative aspects) is crucial. It is in this second area that the professional's values play such an important part. The 'would-be' professional absorbs these values through socialization; the process by which the roles an individual performs and the value orientations required by that role become a part of the individual's personality, reflected in emotional responses and judgements enabling them to be a part of society; or as architects and surveyors, sub-groups of society.

Thus, in the case of professions, one might expect any values to be determined socially through education and training (internalization of roles), and external control (of deviance) exerted by the professional association. For example, in looking at the education and training of professionals, we could analyse the curricula – the body of knowledge the would-be professional acquires – and conclude that since the educators and certifying authorities have only their own experience to guide them, the knowledge will be innately conservative with an emphasis on traditional skills and techniques (e.g. 'taking-off' and 'working up' for quantity surveyors).

The ritualising of these traditional skills creates a 'fellowship of suffering' enabling the probationer to identify with peers during the educational process and with fellow professionals in practice. In all of this, the acquisition of knowledge and of the professional's values cannot be separated. Crucially, failure is always a realistic possibility both in the acquisition of qualifications and surviving in practice: it is fear of failure and its consequence that acts as the spur to continually strive to perfect the service and skills he or she has.

1.4 VALUES IN THE LABOUR PROCESS

Therefore, the professional's knowledge and values are interlinked and determined socially by their training and education. If professionals are separate from other occupations because of this process, we might well expect that they would possess specific and separate values relating to the application of that knowledge as a result.

The traditionally held view is that in western capitalist societies, the process of creating surplus value is the only concern, but such an approach classes value as a commodity and concerns itself only with economic value. From experience it would seem that there is far more to providing a professional service than this because of the different approach to the application of knowledge and the value relationship between the practitioner and their professional activities.

Obviously, as practitioners we have to make money in order to survive, we do not apply this knowledge indiscriminately to make money. Kay's 'good' doctor is motivated by the welfare of his patients and this has the fortunate side effect of making him more financially successful.

Typically, when looking at the labour process, writers like F.W. Taylor look only at the activity, not the person carrying it out. The value relationship between the worker and his or her work is removed from the equation: only the materials and the processes they undergo are of concern in the imposition of rationality on the work situation and to some extent this legacy remains. Commonly there is no mention of the values inherent in the way the work is done, or the service provided; only in its function and exchange (value in use to someone else). This view of the labour and production process may give a clue to the problem of 'alienation': a missing relationship in the work process between the worker and the product of their labour. We would argue that this can be described as the loss of the valuing process that is still inherent in a professional's values.

When political economists studied the dominant technological and economic trends of capitalist organization, they accepted the philosophical approach of the capitalist, ignoring the collapse of a more fundamental value relationship between the worker and the product.

Braverman's 'deskilling' is linked to the loss of control resulting from the loss of knowledge through the process of rationalization. For him, the rationalization causes the alienation, while Carr-Saunders and Wilson note:

> 'Professional men are craftsmen. . . . There is an inescapable desire though it may be latent, to see the fullest and most efficient use made of professional techniques to the services of the community'.

Henry Mintzberg associated the term 'craft' with 'functional art'. The product performs a function but also has aesthetic value. In other words, the pure economic measure of value of capitalism does not apply. He also concludes that one major industry, construction, has also remained largely in the craft stage.

The notion that there might be some underpinning search for excellence, higher values and quality, as opposed to solely making a profit, ties in with the need to avoid failure and the history of the professions stemming from pre-industrial society. For Mintzberg, it is autonomy which 'allows the professionals to perfect their skills, free of interference. They repeat the same complex programs time after time, forever reducing the uncertainty until they get them just about perfect'.

Similarly, Peters and Waterman note that successful organizations often push autonomy downwards by remaining fanatical about primary core values. The socializing process of the profession may work in this way: the autonomy of the individual practitioner stems from core values (ethos and ideology) embodied in the professional institutions (through their codes of conduct) and visible in his or her peers.

If alienation is caused by 'rationalization' then it is understandable to argue that the rationalization of professional work is less advanced than in other non-professional occupations since the value relationship is intact, but it has still occurred. However, if it is carried too far in the organization of professional work and the organization of professional practices, we risk the destruction of the value relationship.

By ensuring that standards are maintained, a professional association ensures that, at entry, the professionals belonging to it have the same basic knowledge, implying that the alienation of the professional does not come from the loss of knowledge but from the standardization and rationalization of the professional's activities.

The professions benefit from a society that values specialized expert knowledge and one of the benefits is autonomy in the execution of their functions. Autonomy does not equal control. The true professional is not controlled by the manager through charisma or rational/legal methods but by professional values (Weber's tradition/historic techniques).

The degree of autonomy which as professionals we believe we possess is an illusion, albeit a comfortable one. However, the trend in the industry now seems to be a shift from this implicit form of control to more explicit forms typified by measures such as Quality Assurance and the new NEC Professional Services Contract.

1.5 RECENT EMPIRICAL RESEARCH

We have looked at knowledge as fundamental to professions, in terms of control in use when passing it on through the indoctrination of their new members. We have also seen that the knowledge, and hence the values in applying that knowledge, would as a result, be common to professionals of the same discipline. However, if the knowledge bestowed differs between disciplines we might expect the values to be noticeably different between professions.

Despite this, we have seen that the client, because of the relationship with the practitioner, needs certain core values regardless of discipline which will reflect similarities in the value relationship between the professional worker and their activities.

A recent pilot questionnaire attempted to identify whether there were noticeable differences and/or similarities in the value relationships between different professionals. The methodology used for the investigation was a questionnaire sent to individual practising professionals of three different construction related disciplines: architecture, quantity surveying and structural engineering. The questionnaire design was based on Powell's ten facets of value (1991) which identified the values likely to be involved in the range of professional activities in a construction project. Respondents were asked to rank the ten values individually according to a semantic scale. The results were subjected to quartile analysis to provide a measure of distribution, the results being as shown in Table 1.1.

Although only a trial survey, the responses indicate that not only are there differences in the relative importance of particular values (for example, Figure 1.2 shows aesthetics to be more important to architects than to quantity surveyors) but more importantly that the distribution of results varies between values. If core values exist, we would expect them to be common to the professions studied and have a smaller range than other 'non-core' values. The quartile analysis shows that this may well be the case. Facets such as 'ethical' show very limited distributions, whereas non-core values such as 'ecological' show wide discrepancies both between professions and samples within professions. Figures 1.1 and 1.2, representing 'ethical' and 'hedonic' values, illustrate this contrast between 'core' and 'non-core' values. The high importance attached to the 'technological' facet reinforces the importance of

Table 1.1 Quartile analysis of value questionnaire

	Ethical	*Aesthetic*	*Technical*	*Prudential*	*Hedonic*	*Ecological*	*Humane*	*Economic*	*Legal*	*Fiduciary*
Architects (sample = 11)										
Q0 (min)	8	7	8	6	4	7	7	6	6	2
Q1	9	8	8.75	7	6.5	7	8.75	8	7.75	7
Q2 (med)	10	9.5	9.5	8	9	8.5	9	9	8.5	8.5
Q3	10	10	10	9	10	9	9.25	10	9	9.25
Q4 (max)	10	10	10	10	10	10	10	10	10	10
Structural engineers (sample = 15)										
Q0 (min)	9	5	8	5	3	2	3	4	4	1
Q1	9	6	9	6	6.75	6.75	5.75	8	6.75	6.75
Q2 (med)	10	8	9	7.5	7.5	8	8	8	8	9
Q3	10	8	10	8	8.25	9	9	9	9	10
Q4 (max)	10	10	10	9	9	10	9	10	10	10
Quantity surveyors (sample = 11)										
Q0 (min)	9	5	4	2	3	5	5	6	2	6
Q1	9	6.75	7.75	6.5	3.75	6.75	5.75	8	6.75	7.75
Q2 (med)	10	7.5	8	7.5	6	8	8	9	8	8.5
Q3	10	8	9	8.25	7	8.25	9	9	9	9
Q4 (max)	10	9	10	9	9	9	9	10	10	10
All professionals (sample = 37)										
Q0 (min)	8	5	4	2	3	2	3	4	2	1
Q1	9	7	8	7	6	7	7	8	7.25	7
Q2 (med)	10	8	9	8	7	8	9	9	8	9
Q3	10	9	10	8	9	9	9	9	9	9
Q4 (max)	10	10	10	10	10	10	10	10	10	10

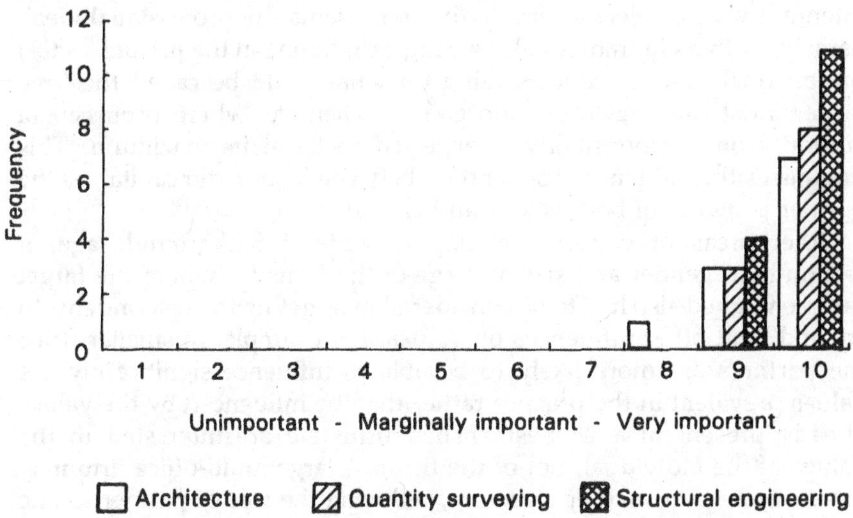

Fig. 1.1 Relative importance of ethical values to construction professionals.

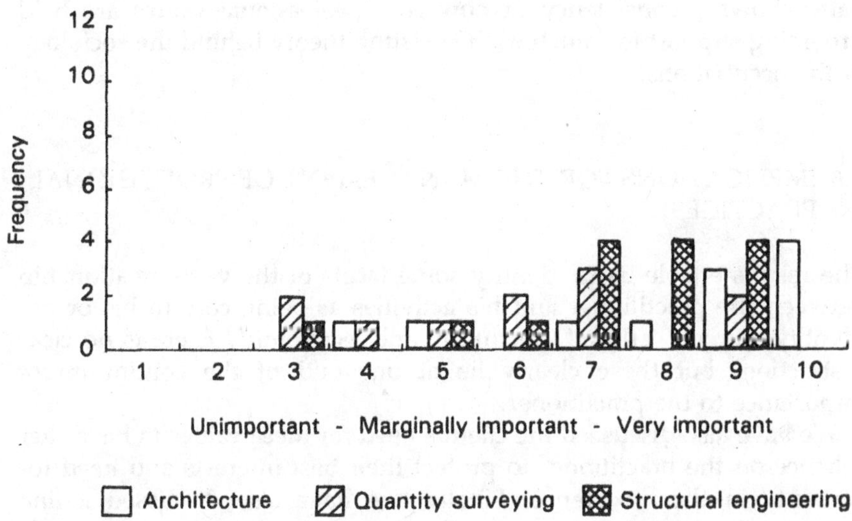

Fig. 1.2 Relative importance of hedonic values to construction professionals.

knowledge at the centre of a professional's identity, while facets such as 'ethical' support the argument for the existence of 'values' in the application of this knowledge.

It must be recognized that in the selection of respondents the intention was to select a sample that represents the professional ideal: namely partners in traditional (i.e. single discipline in the private sector) professional practice concentrating on what could be called the 'free professional catering to an unorganized clientele' where professional authority and responsibility is expected to be at its maximum. This minimizes the influence of separation between labour and capital, as the partner is owner of both labour and capital.

Other areas of concern relating to cultural background, age of respondent, gender and size and age of the firms for whom the target sample worked also had to be considered in targeting the respondents to try and limit other influences on values. For example, in smaller firms the partners are more likely to be able to influence significantly the values prevalent in the practice rather than be influenced by the values already present in a well-established firm (we are interested in the values of the individual, not of the firm). A large multi-office firm may have a more bureaucratic structure, reducing the role of partner to one of manager rather than active practitioner.

Many of the above points have been the subject of research, others are pure conjecture: the intention was not to prove their existence but to limit potential impacts on the results. The data from the trial questionnaire shows a consistency in how core professional values are held providing support for much of the existing theory behind the sociology of the occupations.

1.6 IMPLICATIONS FOR THE MANAGEMENT OF PROFESSIONAL PRACTICES

The results enable us to identify some facets of the value relationship between the practitioner and his activities as being core to his or her identity, namely 'ethical', 'technical' and 'economic'. There is no clear distinction, but these clearly lie at one end of the continuum of importance to the practitioner.

We have also discussed the client's need for these values in his or her reliance on the practitioner to protect their best interests and need for certainty in their existence. Finally, we have also discussed a link between the knowledge of a professional and his or her values – a relationship which could be described as symbiotic in nature.

Together, these three aspects illustrate the differences between 'professional' work and that of other occupations which more closely reflect the 'capitalist' ethos of 'economic' wealth creation. The implications of this for practice management is the danger of treating the supply of 'professional services' as 'just another business'. Time has enabled professions to develop a value structure: a culture of service which allows them to differentiate themselves from other occupations in the same way as Marks & Spencer have succeeded in differentiating themselves from other retailers. It is this value structure that is the professional's greatest asset.

However, Braverman's critique of rationalization – that it leads to a loss of knowledge – implies the loss of these values. One cannot know the best way to apply knowledge in a client's interest without fully possessing that knowledge. Treating professional work as other occupations (automating, standardizing, developing management systems, etc.) all seek to remove knowledge from the practitioner and incorporate it into the company's structures and procedures, undermining the practitioner's ability to make value-judgements and consequently undermining the need for his or her values.

While more and more organizations seek to develop values and acquire 'ethics' in business – seeking the trappings of the professions – professionals in turn must be careful not to find themselves moving in the opposite direction.

1.7 REFERENCES AND FURTHER READING

Braverman, H. (1974) *Labour and Monopoly Capital*, Monthly Review Press, London.

Carr-Saunders, A.M. and Wilson, P.A. (1933) *The Professions*, Oxford University Press, London.

Dalton, G. (1974) *Economic Systems and Society*, Penguin Books, Middlesex.

Dingwall, R. and Lewis, P. (eds) (1983) *The Sociology of the Professions*, Macmillan, London.

Johnson, T.J. (1972) *Professions and Power*, Macmillan, London.

Kay, J. (1995) Economic View; Honesty is the best policy. *The Daily Telegraph*, 20th February.

Larson, M.S. (1977) *The Rise of Professionalism*, University of California Press, Berkley.

Marx, K. (1981) *Capital*, translated by David Fernbach, Penguin Books, Middlesex.

McClelland, D.C. (1976) *The Achieving Society*, Irvington Publishers, New York.

Mintzberg, H. (1983) *Structure in Fives – Designing Effective Organizations*, Prentice-Hall International, Englewood Cliffs, New Jersey.

Moore, W.E. (1970) *The Professions: Roles and Rules*, Russell Sage Foundation, New York.

Morgan, G. (1990) *Organizations in Society*, Macmillan, London.

Peters, T.J. and Waterman, R.H. Jr (1982) *In Search of Excellence*, HarperCollins, New York.

Powell, M. (1991) Axiology Related to Building. *Proceedings of the 7th ARCOM Conference*, University of Bath, pp. 72–84.

Root, D. (1992) *Professional Values*, unpublished MSc Thesis, University of Bath.

Smith, A. (1970) *The Wealth of Nations, Books I–III*, Penguin, London.

Sraffa, P. (1951) *The Works and Correspondence of David Ricardo*, Royal Economic Society.

Taylor, P.W. (1961) *Normative Discourse*, Prentice-Hall, Englewood Cliffs, New Jersey.
Thompson, P. (1983) *The Nature of Work*, Macmillan, London.
Torstendahl, R. and Barrage, M. (eds) (1990) *The Formations of Professions*, Sage, London.

CHAPTER 2

The impact of information technologies on professional advice

2.1 INTRODUCTION

Since the 1980s, information technology (IT) has made a substantial impact in the business world. Managers and professionals have had to come to terms with the redefined skills, needs and demands engendered by the new technology. Furthermore, IT has been identified as a strategic weapon which can improve competitive advantage. To take full advantage of this, however, requires planning for the necessary integration of business and IT strategies. The skills factor is also important in this respect, and the growing importance of human resource management (HRM) and human resource development (HRD) in professional organizations has led to a greater awareness that a successful IT strategy also requires relevant and well-directed strategic IT skills training and education provision.

The property profession is no exception to this, and has also undergone a number of major structural alterations as a result of 'forces for change'. A vital theme running through these is the impact of IT, which has had reciprocal implications for IT skills training and education.

This chapter therefore reviews the theories and debate involving strategic planning and HRM issues for IT, in areas outside property. These are then related to the key focus of the empirical research, the property profession. The results of this research, based on surveys and case studies of IT users, including private practice firms (general practice and quantity surveying), the public sector and corporates, together with IT course providers (the universities and higher education institutions), and carried out by the College of Estate Management, Reading in 1993/94 are described. Conclusions are drawn, and a 'prescription for success', incorporating 'critical success factors' for the property profession to achieve well-integrated IT strategies and effective IT training and education, is developed.

2.2 INFORMATION TECHNOLOGY: THE STRATEGIC AND HUMAN RESOURCE DIMENSIONS

In order to understand how IT has made an impact in the property profession and how the research fits into the context of previous research, it is necessary to examine two main dimensions to IT, which take much of their sustenance from areas outside surveying. These are:

- the strategic dimension; and
- the human resource dimension.

2.2.1 The strategic dimension

Increasingly, the trend is towards an information society where the economy itself is service-oriented. The key resource is information and the enabling technology is IT. Moreover, there is a firmly held belief by many that IT is a strategic resource because it brings about or facilitates fundamental changes in industry sectors, in competitive behaviour, and in an organization's own strategy, structure and functioning. As Earl points out, in its move from a 'data processing' era to an 'end-user computing' era, IT has been used as a strategic weapon in four main ways:

1. To gain competitive advantage.
2. To improve productivity and performance.
3. To enable new ways of managing and organizing.
4. To develop new businesses.

The key to using IT for competitive advantage is to direct and target IT into the product and service, into the channels of distribution and supply, or to change the basis of competition against rivals.

But what is a 'strategy'? There is certainly no agreed definition of this term, and Mintzberg recognizes this by presenting five definitions of strategy: as plan, ploy, pattern, position and perspective. Glueck believes that:

> Strategy is a unified, comprehensive and integrated plan . . . designed to ensure that the basic objectives of the enterprise are achieved.

This reinforces Mintzberg's idea of strategy as 'plan', which is a consciously intended course of action, or set of guidelines to deal with a situation.

If this is accepted, the question is what should comprise an IT strategy? Earl suggests three components (Figure 2.1): systems, technology and management. The 'systems' component is concerned with determining the applications to be computerized; 'technology', with

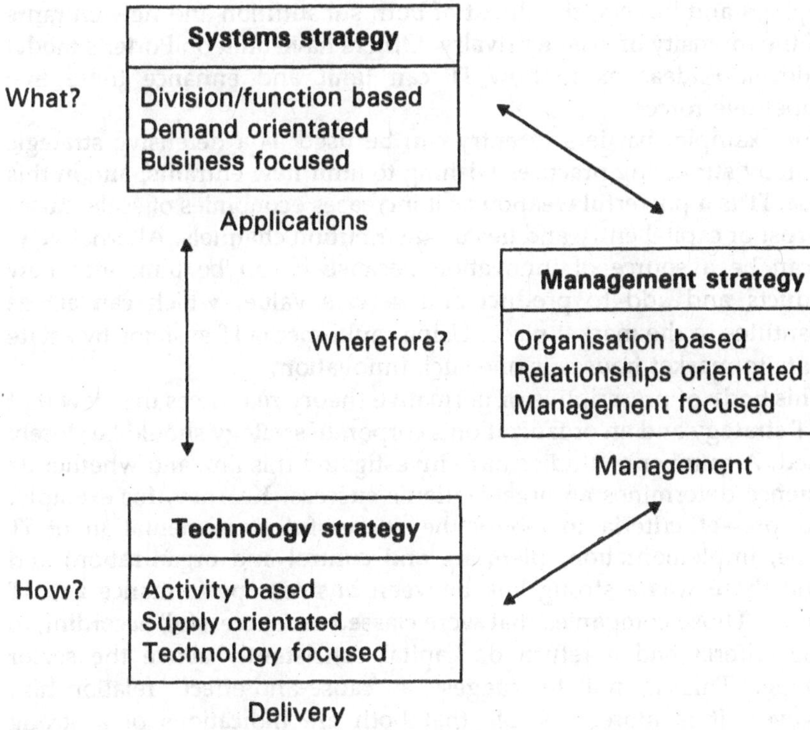

Fig. 2.1 Components of IT strategy for firms.

how the strategy may be delivered, and 'management' with the policies (including IT training and education), procedures, aims and actions needed to implement the strategy.

Clearly, IT is a valuable strategic weapon, although the application of a strategic perspective to IT is a relatively recent phenomenon. A number of frameworks have been developed however, which are designed to emphasize the 'technology–strategy' connection so that organizations can exploit IT for strategic advantage.

One group of well-documented and respected prescriptive models to achieve a better understanding of this relationship is known as 'opportunity frameworks'. These are analytical tools designed to lead to specific strategic opportunities and clarify business strategies to demonstrate options for using IT in a strategic role. Porter's model of competition and industry analysis is one such example. Building on linkages between management policy and individual organization he proposed a framework of five forces which defined the basis of competition within an industry: the bargaining power of both existing suppliers and buyers, the threat of both substitution and new entrants and the intensity of existing rivalry. Others have built on Porter's model to develop ideas as to how IT can limit and enhance these five competitive forces.

For example, barriers to entry can be used as a defensive strategic action by surveying practices wishing to limit new entrants, and in this sense, IT is a powerful weapon as it increases economies of scale, raises the cost of capital entry and ties up distribution channels. Alternatively, IT can be a source of innovation because it can be built into new products and add to product and service value, which can act as substitutes in the market place. Using multi-media IT systems by estate agents to market houses is one such innovation.

This body of prescriptive or normative theory reinforces the view that an IT strategy and an organization's corporate strategy should be closely linked. A number of studies have investigated this link and whether its existence determines an organization's success. Kearney, for example, used pre-set criteria to assess the successful implementation of IT (scope, implementation, planning and control and organization) and found there was a strong link between business performance and IT success. Those companies that were classed as 'successful' according to these criteria had a return on capital consistently above the sector average. This is not to suggest a 'cause-and-effect' relationship, however. It is more probable that both are indications of a strong management team that is as successful in the use of IT as in other areas of managing the business. This research also found a strong link between 'successful users' and their recognition of the relationship between business and IT strategy: 40% of successful users developed their IT strategy from the business plan, but also let the latter be influenced by the former. Furthermore, the importance of structure in an organization was recognized: strategy needs to be evolved within the context of structure and culture to be able to detect, and respond dynamically, to changes in the market place.

This idea of 'strategy and structure' was further extended by Peters and Waterman in their '7-S' framework. Based on previous work by Leavitt, their model incorporated seven interdependent variables which they believed were essential for organising intelligently: structure, strategy, people, management style, systems and procedures, culture and strengths and skills.

2.2.2 The human resource dimension

Skills, or the 'people factor', is vital to consider in relation to any strategy. In this sense, training is seen as part of the quality process which should be fully integrated, like IT with business strategy. This viewpoint has given rise to a philosophy known as human resource management (HRM) and provides the second of the two dimensions to IT described earlier in this chapter.

Armstrong describes HRM as:

> A strategic, coherent and comprehensive approach to the management and development of the organization's human resources, in which every aspect of the process is wholly integrated with the overall management of the organization.

HRM is therefore an ideology which is underpinned by the premise that organizations exist to deliver value to their customers and that this is best achieved by adopting a longer-term perspective to managing people, and treating them as assets.

At the lower level, and as a result of HRM, human resource development (HRD) is vital to consider as it is concerned with enhancing and widening IT and other skills through education and training. Such reasoning is not without its critics, however. It is recognized, for example, that it is extremely difficult to deal with HRM in a strategic sense and to align business and human resource strategy: this may not be possible because of the diversity of strategic processes, levels and styles in an organization, for example.

Nonetheless, research by the Training Agency indicated that HRM and related HRD could go a long way to resolving the problem of the 'skills gap' that existed in the UK. Such a gap exists at a national level in terms of vocational education and training; in management training and education, and in IT skills training and education. The British Institute of Management research, for example, suggested management is receiving inadequate and inappropriate IT skills training. Few organizations planned or managed IT training and education and there was poor integration of IT with corporate strategy. Only 36% of respondents in the BIM study for example, claimed their strategies were well integrated. Moreover, IT training and education was seen as falling between two stools: it was not recognized at a strategic level and was often only dealt with at an operational level. The impression is that the 'management' component of the IT strategy is not catered for sufficiently. Can the same be said in the property profession?

2.3 THE IMPACT OF IT IN THE PROPERTY PROFESSION: AN OVERVIEW

As we have described elsewhere, the property profession has faced a number of important forces for change over the last few years. These comprise:

- external competition;
- changing client demands;

- technological change; and
- internationalization and globalization

External competition has been driven by the impetus of financial deregulation, which has led to other professionals such as lawyers and accountants, together with management consultants offering a 'one-stop' consultancy service. In the same way, clients have become more sophisticated during the 1980s and 1990s in the advice they seek, and the perception of value for money that they place on professional advice has also been heightened. Technological change itself is also important. This has led to a shift towards the high ground of management in surveying. Former professional tasks can now be described as technical work as a result of the impact of IT. Finally, internationalization and globalization have redefined the environment in which the property profession finds itself. To take one example, global property investment by global managers based in different countries has become commonplace, and by being international and multi-capacity with high-quality service reputations they are already in a position to make in-roads into general practice surveying firms' markets.

This is not the place to discuss the applications of IT within the property profession. Indeed, applications or potential applications are one thing, hard evidence of the level of use of IT systems is another. Recent surveys suggest levels of usage for IT are relatively high in the profession: in percentage terms, 80–90% of surveying practices/organizations use technology to automate all or part of their standard procedures.

What, though, is the effectiveness of such usage? The important factors for this are, clearly, strategic planning for IT and a fully integrated IT training and education programme (the 'management' component of the IT strategy). The evidence for the presence of these is, however, disappointing. For example, one study found that a strategic plan for IT only existed in a minority (21%) of the sample of surveying organizations.

There has also been much debate as to where IT fits into the skills base of the profession. In 1992, for example, the Moohan report described IT as a 'medium order' skill (useful or desirable in some way). Only in relation to land surveying was IT ranked as a 'high order' skill. In contrast, the RICS in 1989 saw IT as a 'core content' skill, but the RICS QS Division Report in 1992 described IT as a 'tool'. The conclusion cannot be escaped: IT is important but the profession as a whole appears to be uncertain as to **how** important. This cannot be good for the property profession given that IT is such an important force for change in the general practice and quantity surveying divisions.

There is clearly uncertainty over the exact role of IT skills. Moreover, there are other related research questions which need to be answered including:

- What is the level of strategic planning for IT in the surveying profession?
- How effective is IT training and education strategy in the profession?
- What are the levels of ability of professional surveying staff in IT skills?

The College of Estate Management's (CEM) research, conducted in 1993 and 1994, was designed to answer these and other questions (for details, see Dixon, 1995).

2.4 EMPIRICAL RESEARCH: METHODOLOGY

The research study comprised a postal questionnaire survey of senior and associate practitioners (general practice and quantity surveying) in 400 organizations, to which 132 replied. The sample (and subsequent responses) included an equal weighting of private practice firms, the public sector and corporates (including commercial and financial service companies). In addition, all universities and other higher education institutes which deliver GP and QS surveying courses, were sent a questionnaire: 19 responses were received out of a total of 36 questionnaires despatched. These were supported by four in-depth case studies of two large GP firms, a medium-sized QS firm and a corporate organization.

2.4.1 Summary of results

Strategic planning for IT

The majority of respondents (75%) considered IT to be 'essential' to their organizations. This was less so with respondents in smaller organizations but more so in corporate organizations. Most organizations (77%) had used PCs and other forms of IT for more than 5 years.

The majority (80%) of the respondents had some form of business strategy, but this was not true of the public sector. Although some 57% of respondents had a full, formal IT strategy, 23% had a partial formal strategy, 6% an informal strategy and 15% had no strategy at all. Moreover, 30% of small organizations had no IT strategy; only 28% had a full, formal IT strategy. This seems to confirm the findings of Sullivan whose sample was heavily weighted towards small organizations (some 73% of the sample had less than 20 employees). His study showed only 21% had a formal IT strategy; but among larger organizations the average was 51%.

Interestingly, the CEM research found that corporates, general practice and quantity surveying organizations all have a greater tendency to incorporate an IT strategy than the public sector. The type of IT strategy implemented also varied: smaller organizations tended to have 'laissez-faire' IT strategies, lacking central control and guidelines; larger ones had been able to evolve high-level, 'marketing' and 'operations' IT strategies.

The 'management' component in the IT strategy, which includes IT training and education, was the least common component. This is of concern because such a component should be dealing with the aims, policies and actions needed to implement the IT strategy.

Consultation with staff, however, was normally carried out by many organizations, although 30% did not. The GP and QS surveying practices appear to be better at this than other types of organization.

Of those with some form of IT strategy, about 75% of respondent organizations had also integrated the IT and business strategies well, in the opinion of respondents. Corporates were best at doing this; the public sector, with fewer business plans, were much less successful. Comments from respondents suggested that full consultation with staff, and a business plan which drives the IT strategy, are among those factors critical to success. Failure to integrate was marked by a piecemeal approach, lack of consultation and lack of support.

In terms of justifying IT, the maximization of office productivity and customer service requirements were considered to be two of the most important strategic factors. Competitive advantage was very important for surveying practices.

In the post-implementation period, delays in implementation and significantly, lack of training for end users, were two problems which received most attention.

IT training and education strategy

As far as assessment of needs is concerned, self-selection predominated, although annual appraisal was also common. However, 17% carried out no formal analysis at all, particularly in smaller organizations, when staff preferred to deal with training needs on an 'as-and-when' basis.

Training levels were lower in IT in comparison with other skills, and were characterized by informal or self-teaching modes of delivery. In fact, a majority (59%) felt appropriate IT training and education had not been provided by their organizations. Although the provision of IT skills by universities on relevant courses was found to match the pattern established in the profession, evidence existed of a skills gap or a shortage of practitioners (especially older members) with the right IT skills, particularly in relation to planning, managing and using IT strategically. This was evidenced by a gap between 'ability' and 'importance' ratings for various IT skills tasks.

Differences within the profession

It should be stressed that there are noticeable differences when the respondents are broken down into sub-groups. Corporate organizations, for example, were the most sophisticated and au fait with the strategic implications of IT, and most considered IT to be essential to the running of their business. The GP and QS surveying practices were fairly close in their characteristics: both groups were mature users of IT, and competitive advantage and other strategic factors were important in justifying the use of IT.

2.5 A PRESCRIPTION FOR SUCCESS

Despite some positive features from the research findings the fact remains that the level of strategic planning for IT in the profession is low and there is evidence of a 'skills gap' among practitioners (especially older members). The case studies which supported the postal questionnaires identified critical success factors which could promote the effective provision of IT training and education in the profession. A number of hypotheses were also tested statistically to investigate the most important factors affecting IT training and education at organizational (the organization itself) and individual (the employees within the organization) levels.

Two groups of factors were found to be important in this part of the research: 'intrinsic factors', such as size, type of organization, and maturity of use, and 'critical success factors', such as importance of IT use and culture. The term 'intrinsic' is intended to describe factors, which although emanating from within an organization, are relatively fixed in the short run. 'Critical success factors', on the other hand, are easier to implement or change. Culture can be changed through committed 'champions for change', staff can become more fully involved in decision making and so on. The key point is that successful strategic planning for IT can be achieved by changing these factors so that the intrinsic constraints of size and organization may be reduced.

This is shown in Figure 2.2. For example, at an organizational level a formal IT strategy, which is well-integrated with a business strategy requires:

- an organization which attaches importance to IT;
- staff who are fully involved in the strategy; and
- a strong management culture.

Fully integrated with this should be an IT training and education strategy which itself requires:

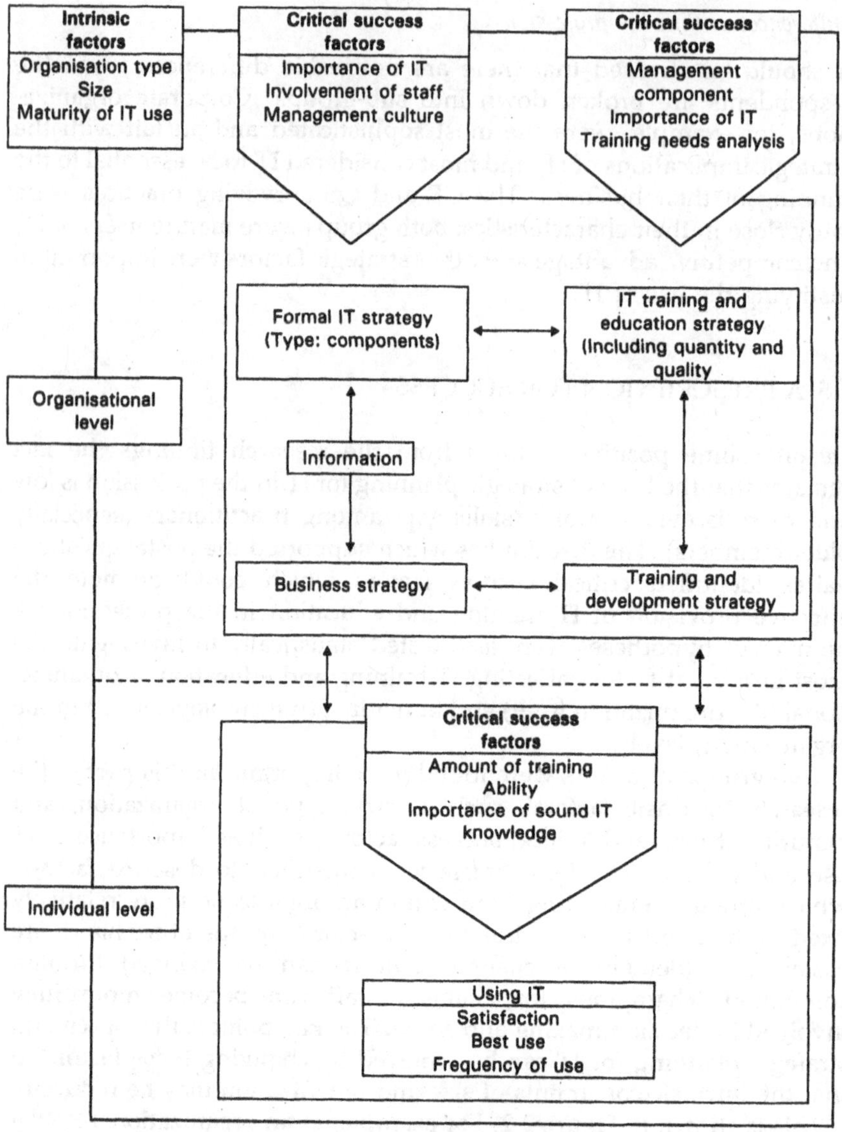

Fig. 2.2 Successful IT skills training and education in surveying firms.

- a management component within the IT strategy to take care of education and training;
- a 'pro-IT' organization; and
- training needs analysis to determine training and education requirements.

At an individual level, individuals are more likely to be satisfied, make the best use of IT and use IT more frequently if they:

- receive adequate training;
- are able users; and
- attach importance to having a sound IT knowledge.

These critical success factors operate within an 'intrinsic factor' framework which can either constrain or drive the former factors.

2.6 IMPLICATIONS OF THE RESEARCH FOR THE PROPERTY PROFESSION

The research has shown there is a need to:

- introduce more effective strategic planning for IT;
- improve IT training and education at an operational level; and
- promote the education of senior managers in matters relating to the strategic use and value of IT.

Effectiveness at the level of strategic planning and integrated training and education is often lacking, and dissatisfaction with the provision of such training and IT skills levels is widespread. Given the changes which are taking place in the property profession, such as the 'flattening' of organizations, the increasing use of IT in professional work, and a greater emphasis on the management role in surveying practices, these shortcomings are disturbing.

IT is having a major impact in the property profession yet in many instances managers are ill-equipped to use and manage IT. The 'skills gap' which is present is not a short-term, recession-led gap: other types of training have not suffered to the same extent. The place of IT in the professions' skills base remains uncertain: perhaps this is a structural problem with many older professionals remaining unconvinced and resistant to IT.

In the past, the universities have an important role to play, not only in meeting the future needs of new entrants but also in providing IT courses relevant to the needs of existing professionals. The task in hand is to plan strategically for IT and improve the IT training and education of all those who need it in the property profession.

2.7 REFERENCES AND FURTHER READING

Armstrong,N. (1992) *Human Resource Management: Strategy and Action*, Kogan Page.

British Institute of Management (BIM) (1988) *Managers and IT Competence*, BIM.

Cash, J.I and Konsynski, B.R. (1985) IS Redraws Competitive Boundaries. *Harvard Business Review*, March–April, 134–42.
CICA/KPMG (1993) *Building on IT of Quality*. Construction Industry Computing Association (CICA/KPMG), Peat Marwick, Cambridge.
Dixon, T.J. (1995) IT Skills Training and Education for the Surveying Profession: Requirements for the 1990s. College of Estate Management Research Paper 1995/1, CEM.
Earl, M.J. (1989) *Management Strategies for Information Technology*, Prentice-Hall.
Glueck, W.F. (1980) *Business Policy and Strategic Management*, McGraw-Hill.
Handy, C. and Gow, I. *et al*. (1987) *The Making of Managers*, National Economic Development Office (NEDO), London.
Kearney, A.T. (Management Consultants) (1990) *Breaking the Barriers – IT Effectiveness in Great Britain and Ireland*, A.T. Kearney and Chartered Institute of Management Accountants, London.
Leavitt, H.J. (1978) *Managerial Psychology*, 4th edn, University of Chicago Press.
McFarlan, F.W. (1984) Information Technology – change the way you compete, *Harvard Business Review*, May–June, 98–103.
Mintzberg, H. (1987) Five Ps for strategy, *California Management Review*, Fall, 80–8.
Moohan, J. (1992) *An Analysis and Evaluation of the Education and Professional Practice Integrity of Fully Exempting Post-Graduate Conversion Courses*, RICS.
Peters, T.J. and Waterman, R.H. Jr (1982) *In Search of Excellence: Lessons from America's Best-Run Companies*, Harper Collins.
Porter, M.E. (1980) *Competitive Strategy*, Free Press, New York.
Porter, M.E. and Millar, V.E. (1985) How information gives you competitive advantage. *Harvard Business Review*, July–August, 149–60.
RICS (1989) *Future Education and Training Policies – Report of the Education and Membership Committee to the General Council*, RICS.
RICS/QS Division Research Group (1992) The core skills and knowledge base of the quantity surveyor. *RICS Research Paper* 19, RICS.
Sullivan, K. (1993) *The Strategic Use of IT (RICS Survey)*, South Bank University/ RICS.
Training Agency (1989) *Training in Britain: A Study of Funding, Activity and Attitudes*, Summary Report, HMSO, London.

CHAPTER 3

The technical professional as a management consultant

3.1 INTRODUCTION

The 1990s have brought many changes which have had a significant impact on private practice property professionals. The property recession has led to a decline in many of the traditional areas of work, particularly those related to transactional business and new build projects. At the same time, the core client base of many private practices – organizations for which property is a core activity such as developers, investors and construction companies – were particularly affected by the general economic downturn.

Organizations using property as a resource to support their core activity faced a different situation. In order to react to the recession, many had to find ways of cutting costs. This led to a reduction in the workforce which brought with it a realization that organizations are committed to the significant cost of accommodation which cannot be reduced in quite the same way. For many decision makers this was the first time that property had become a significant issue on the business agenda.

Major occupiers required property-related management advice, and a market for the providers of property services, that had remained untapped for years, began to emerge. A number of surveying firms have recognized this market and are providing new services to meet the needs of this group. The territory is new and requires a different approach and different skills.

This chapter examines this changing market place and discusses how a different approach is needed to meet the needs of occupiers. The shifting role from technical specialist to business adviser has an impact not only on the skill and knowledge base required but also on how firms are organized to respond to that market.

3.2 THE CHANGING ENVIRONMENT

3.2.1 The business sector

The commercial environment has changed rapidly in the 1990s. All types and sizes of organizations are now facing the challenge of surviving in a highly competitive environment. In order to succeed, organizations are turning to new management initiatives to preserve and improve their profitability by reducing expenditure while at the same time retaining their market share and their income. Business process re-engineering, delayering, downsizing, flexible working, out-sourcing and quality improvement, are the current management buzzwords and techniques being used by firms to meet this challenge.

In the 1980s increased profits generally came from increased turnover as the economy expanded and demand increased. The need to consider the expenditure side of the budget was important but not paramount. As the recession has bitten hard into profit, all organizational resources have come under scrutiny. Human resources were the first and most notable to be subject to the changing attitude of the Boardroom and now it is the turn of other resources to come under the microscope. Among this second wave of corporate cost auditing is the property resource and, perhaps surprisingly, the particular interest in property is not in its value but in its cost in use. Organizations are recognizing that to be successful, the link between effective resource management and long-term performance must be acknowledged.

3.2.2 The public sector

It is not only in the private sector that the containment of expenditure is leading the management of organizations. In the public sector, current levels of public expenditure as a percentage of gross domestic product is increasingly being considered by governments of many developed countries to be unacceptably high; an increasing 'dependency ratio' (the ratio of those who are economically active to those dependent, in one form or another, on government expenditure) will progressively burden government; government's desire to reduce the large Public Sector Borrowing Requirement (PSBR); and technology's ability to deliver healthcare and public services in excess of our ability to pay for them; are all combining to stretch the finance available for government to spend. In parallel with the business sector the situation requires a similar containment of expenditure.

3.2.3 Markets

This is not all that is changing. It is not only the recession that has caused this fundamental rethink in corporate and public sector life.

Technology generally – and information technology in particular – is changing the way that markets react and the way organizations do business. Markets have become global and changes in local markets have become erratic and volatile. Organizations must exercise very deft footwork to retain their sharp market focus and this too is changing the way they think and respond.

The key characteristics of successful organizations in the future are likely to be responsiveness, flexibility, and adaptability with a global outlook.

3.3 A VIEW OF THE FUTURE

3.3.1 The new commercial environment

The consequences of this view of the commercial environment will be to produce organizations with flatter structures with a task-based work culture. Matrix management will replace traditional hierarchies, while working practices will become increasingly flexible as hot desking, hotelling and teleworking develop. Organizations will have a strong customer focus, driven by customer requirements as opposed to being driven by the requirements of administering their own business. In order to simply stay in the market, companies will have to reduce cost and improve quality and service rapidly and continuously.

3.3.2 The impact on property as a resource

So how do these changes manifest themselves in organizations and how is this affecting property?

The first and most obvious change is that all property and facilities costs are coming under scrutiny. There is an awareness that property has been under-managed in the past and consequently there is a perception that costs have been uncontrolled. Any reasonable cost savings will come 'straight off the bottom line'.

Second, organizations are preserving or improving their profits, while at the same time they are downsizing with a consequent reduction in their property requirements. They are growing and getting smaller! This leads to organizations wishing to divest their surplus property into an already very weak property market. Similar changes are occurring in the public sector.

Third, organizations are seeking more flexible buildings and lease terms as the business planning cycle shortens and organizations realize that their accommodation requirement will change with increasing rapidity.

Fourth, new working practices are gradually changing organizations' requirements of the configuration of their accommodation. More team working space and discussion space is required and a better working environment to encourage more flexible and innovative thinking and creativity. At the same time, less desk space is required as hot desking and out-of-office working increases.

These changes suggest a number of consequences for property, not least a shift in ideology, as property is considered as a key resource for organizational management and development, and not simply an asset. Figure 3.1 illustrates the component parts of the property resource; Figure 3.2 indicates the key management issues that impact on property. These changes also suggest that overall the demand for space will fall, and the remaining space requirement will be for flexibility in both physical and lease terms. Business management and property management decision making will become linked and on an operational level, cost control will incorporate building maintenance, energy efficiency and facilities management. Public sector and business sector property

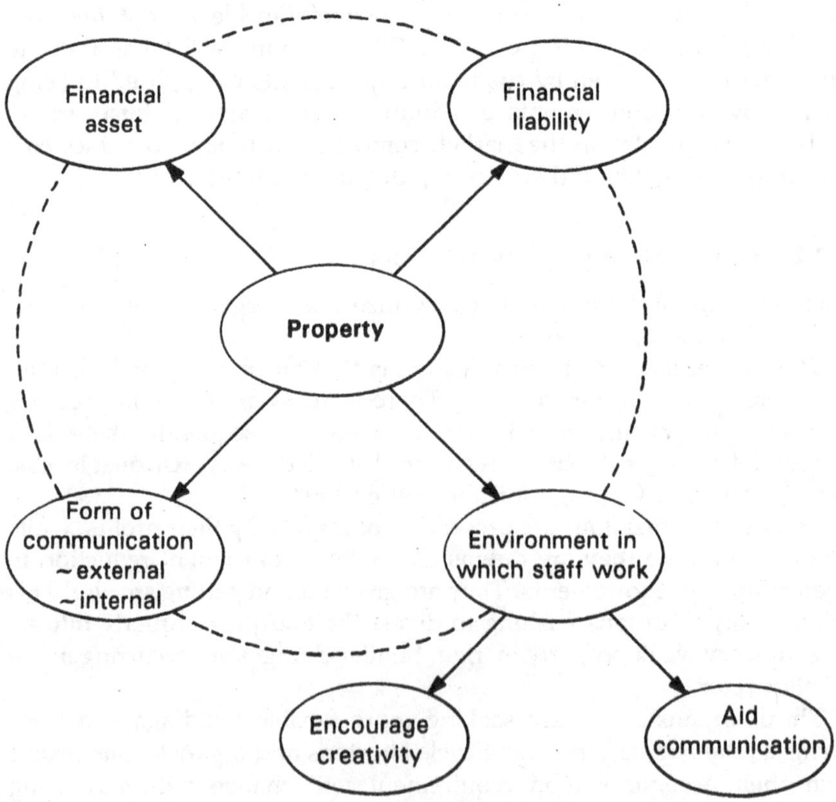

Fig. 3.1 Component parts of the property resource.

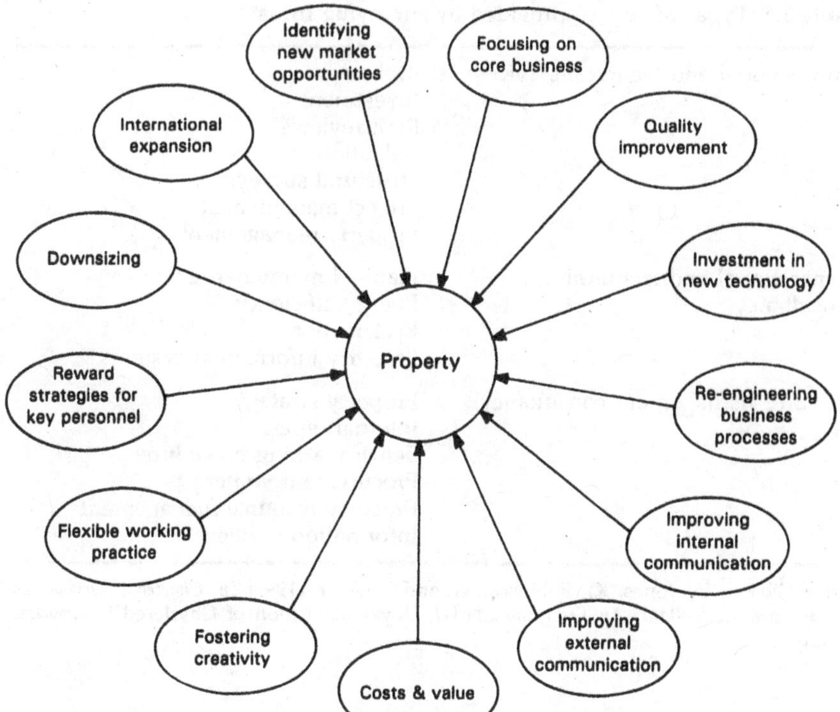

Fig. 3.2 Key management issues which impact on property.

will follow parallel lines with the obscuring of previously clearly defined boundaries between the two sectors.

3.4 THE TRADITIONAL FOCUS OF THE CHARTERED SURVEYOR

The main question is, where does this leave the chartered surveyor and chartered surveying firms? This new emphasis on the property resource by major occupiers, is likely to mean that the property sector – traditionally the focus of the general practice surveying profession – is likely to remain static or experience decline, as contracting demand for space may mean a slow reduction in transactional and institutional investment business.

The traditional functional division of services within general practice surveying firms is unlikely to meet the needs of the expanding occupier

Table 3.1 Types of service provided by surveying firms*

Transactional and technical services	Brokerage Investment Rent reviews Valuation Structural surveys Project management Property management
Transactional and technical consultancy	Planned maintenance Energy efficiency Reclamation Property information systems
Resource management consultancy	Property strategy Internal rents Bench-marking procedures Procurement strategy Property resource management Information systems

*From Gibson, V., Jones, K., Robinson, G. and Smart, J. (1995) *The Chartered Surveyor as Management Consultant: An Emerging Market*, Royal Institution of Chartered Surveyors, London.

sector, as occupiers no longer view their property needs as compartmentalized into disparate transactional and technical areas, such as rent reviews, lease renewals, acquisitions and maintenance. A more comprehensive list of property services classified in three main groups: (i) technical and transactional Services; (ii) transactional and technical consultancy; and (iii) resource management consultancy (Table 3.1).

Table 3.1 provides a framework for classifying the services that a chartered surveyor delivers.

The traditional focus has not seen Property Resource Management (PRM) consulting as a significant business opportunity, but a declining demand in the traditional areas of property services calls for creativity and innovation in approaches to securing work and meeting customer requirements in these expanding markets of the future.

3.5 THE CHALLENGE OF NEW OPPORTUNITIES FOR SURVEYORS

In order to meet the challenges of this expanding PRM market, general practice surveyors and property advisory firms must look beyond property and gain a real understanding of the business they service. This will enable the consideration of property in the wider context, and not simply technical solutions isolated from other relevant business issues. This new form of wider advice will develop at both strategic and operational levels within the occupier sector.

At the strategic level this will require greater research into the nature and dynamics of the property market, and of the occupier sector in particular. In simple terms, the general practice surveyor will need to understand what outputs are required from occupiers' property systems.

On an operational level, the key issues for the property occupier relate to cost reduction, including lower running costs, disposal of surplus space and improved energy efficiency. This type of active management also includes increasing the flexibility of space, reducing rental costs, improving standards of maintenance and repair, and acquiring new accommodation, and will probably also reflect the gradual merging of the market place for property and facilities services.

In order to resolve these issues, occupiers are seeking advice on a number of techniques and processes splitting corporate and occupier functions, developing occupier agreements and internal charging, together with central and devolved property and accommodation responsibilities. There is also an emphasis on the management of property-related information especially in terms of the development of performance indicators and bench-marking.

This new realm of property advice requires the general practice surveyor's normal 'technical tool kit', alongside knowledge of management and business processes, and finance and accounting. The development of this market will involve some individuals with this wide range of management skills while others will develop in-depth, but narrower, property skills. In order to combine these skills it is likely that property advisers will be working in multi-disciplinary teams in order to tackle management problems and opportunities as a whole, similar to the way in which management consultancy practices operate. Property advisory firms face similar challenges and it is important that firms develop a strong customer focus, linked to internal organizational changes, such as a single point of contact for their clients, and the use of multi-discipline teams and they should not rely upon a product or service focus to expand their business. Organizations will want their property advisers to fully understanding the dynamics of their own business and participate directly in the decision-making process. Figure 3.3 shows the chartered surveyor entering the decision-making process. In essence, this is a shift from technical adviser to business consultant, and the forming of a strategic alliance between property adviser and client is likely to develop into a partnership approach, and not isolated advice relating to a single issue. Similarly, the devolving of the responsibility of management and property through outsourcing will require a new set of skills and knowledge which property professionals may not have needed in the past.

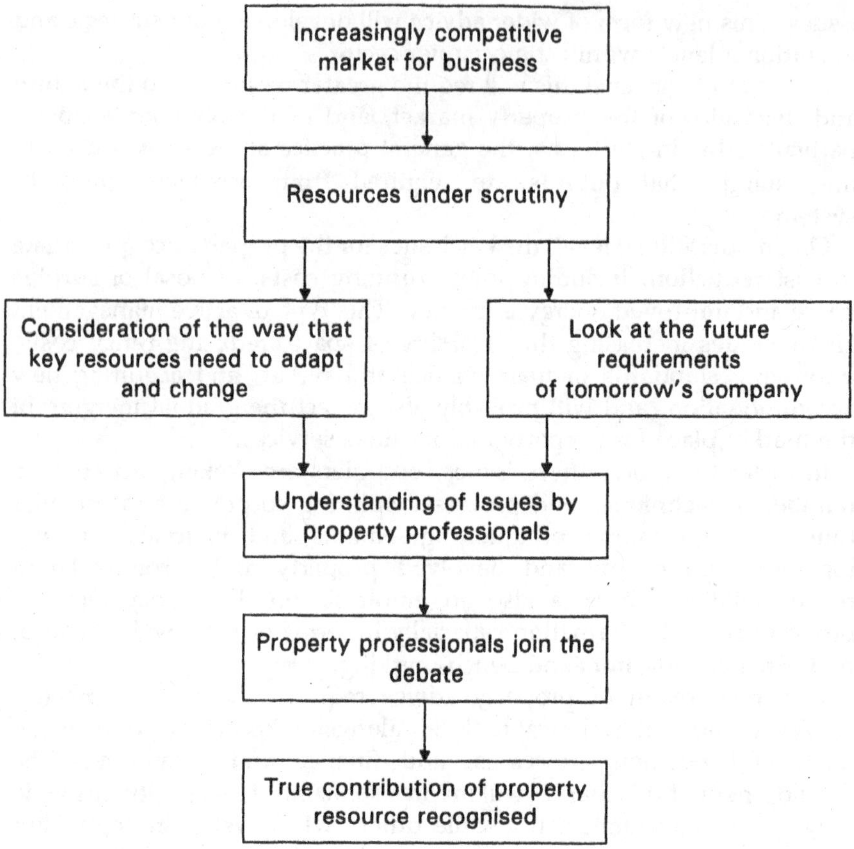

Fig. 3.3 How property professionals may enter the organizational decision-making debate.

3.6 CONCLUSIONS

Property professionals who provide advice – either externally or internally – on operational property management must meet a number of expectations. Occupiers will expect professionals to fully understand their business and to participate fully in the decision-making process. There is also evidence to suggest that the landlord sector is also becoming more sophisticated in its requirement for property advice. Their need is for a better understanding of business objectives, balance sheets, profit and loss accounts, and management techniques.

This changed market place requires a combination of skills, combining the traditional property skills of the chartered surveyor with problem-solving flair.

Firms in the property sector are generally run by property professionals who, in many cases, have little management training. This emerging market presents these professionals with the challenge of examining the way that their organizations are managed. This examination may encompass taking a strategic view of their markets, considering their current structure and reviewing their existing working practices. This process is vital if organizations are not to be marginalized by potential competitors.

The majority of property professionals will find that these wider skills represent a new challenge and that the traditional image of the general practice surveyor as a technical expert and not as a management adviser has hitherto been a barrier to entering this expanding market.

A plan of development is necessary in order to promote the idea that there is a connection between the work of general practice chartered surveyors and more general business advice, and that the combination of the two can add significant value to chartered surveyors' clients.

3.7 REFERENCES AND FURTHER READING

Avis, M., Gibson, V. and Watts, J. (1989) *Managing Operational Property Assets*, The University of Reading, Reading.

Debenham Tewson Chinnocks (1992) *The Role of Property: Managing Costs and Releasing Value*, DTC Research, London.

Institute of Management (1994) *Management Development to the Millennium: The New Challenges*, Institute of Management, London.

Graham Bannock and Partners (1994) *Property in the Boardroom*, Hillier Parker, London.

Gibson, V.A. (1995a) Is property on the strategic agenda? *Property Review*, **5**(4), 104–9.

Gibson, V.A. (1995b) Managing in the millennium, *Estates Gazette*, **9518**, 109–12.

Gibson, V. Jones, K. Robinson, G. and Smart, J. (1995) *The Chartered Surveyor as Management Consultant: An Emerging Market*, Royal Institution of Chartered Surveyors, London.

RSA (1994) *Tomorrow's Company: The Role of Business in a Changing World*, Royal Society for the Encouragement of Arts, Manufacturers & Commerce, London.

Watkins, J., Drury, L., Preddy, D. *et al.* (1992) *From Evolution to Revolution: The pressures on professional life in the 1990s*, University of Bristol, Bristol.

CHAPTER 4

A new management theory for professional organizations

4.1 INTRODUCTION

In the past, management theory has failed to adequately recognize the qualitative differences between the organization and management of a professional private practice and that of a purely business enterprise. The aim of this chapter is to use the surveying profession as a case study in order to examine the ideology of professionalism. The point under discussion is what do firms believe are the basic characteristics which make them 'professional'. The supposition is that the ideology of professional firms:

- is different from that of the general firms recognized in traditional management theories; and
- has come under increasing pressure to change (and perhaps change permanently) in the current recession.

In developing the argument that surveying firms demand new theoretical ideas, it is possible to create three models. These are:

1. The traditional professional firm.
2. The flexible professional firm.
3. The new professional (post-recession) firm.

Surveying is a profession. It is controlled by a professional body (The Royal Institution of Chartered Surveyors), has regulations and rules of conduct, minimum entry requirements and a recognized career progression in order to obtain professional qualifications. Surveyors, in being professionals, obey its strictures and act professionally. Surveying firms act in a professional manner, being both professional firms and employers of professionals.

Professional firms are faced with twin obligations; those of professional integrity and those of the customer and market forces. Traditionally, the two obligations have been achieved together, since professional firms guarantee professional work and thereby quality and general customer satisfaction. For individual surveyors, professionalism has resulted in carefully designed career routes with assistance and understanding from their employers; after all, notions of firm professionalism depends upon employed professionals who are obeying externally laid down strictures of professional conduct. This tradition of professionalism has led to the development of the 'Traditional Professional' firm.

4.2 THE TRADITIONAL PROFESSIONAL FIRM

The surveying industry has been traditionally portrayed as consisting of a plethora of small firms. The firm is small, with a few partners who see themselves not only as responsible for the firm itself, but also for its employees via education and training, flexible working, responsibility, social and welfare interests.

Surveying firms can thus be portrayed as a unique organizational subset of the typical firm (business unit) represented in traditional organization literature. Surveying firms have evolved over a long period of time within a strong professional history, and have retained much of the fundamental essence of these times. This history can be seen by the traditional reliance on Victorian business organization (partnerships) and being tied to that period's strong views on professionalism and obligation. In modern times, the ideal of both the partnership and the professional has changed, and while it does not necessarily follow that firms are old-fashioned, surveying still retains many relics of the past.

It is for this reason that the small surveying firm in some way retains in its very nature those Victorian ideals, and is traditionally regarded as typical for surveying firms. The term 'small' in this context is used in comparative terms. Broadly speaking, it represents the sole trader/partnership legal structure, and operates on both flexible and *ad hoc* organizational principles. Avis and Gibson (1987) say:

> . . . firms with less than 10 full time employees are considered to be small organizations. . . . The activities of this group cover the entire range of surveying services. The majority of the individual firms have identified a niche in the market and are concentrating on a very limited range of services and/or clients. . . . Small GP firms are better able to adapt to changes in the market because of their size. Their management styles tend to be much more flexible and informal. The success of the firm is dependent on the professional abilities of the partners.

In surveying, like other Victorian professions, ideological emphasis was placed on service rather than profits and on quality, rather than on the scale economies that could be derived from standardized high-volume turnover. Ideals of professionalism are fundamentally delineated within this heritage. Traditional definitions of professions which recognize them as a specialized authority based upon a superior knowledge of a body of theory, governed by controlled entry within a code of ethics and a monitored self-discipline, can only be applied within this Victorian framework. The surveying profession cannot be separated from its own professional culture, which is firmly rooted within the culture of Victorian Britain. Any definition of surveying professionalism must recognize the essential status of this culture in the workings of that professionalism within the firm. Ball acknowledges this when he identifies the typical duties of the quantity surveying firm:

> When quantity surveyors perform independent advisory roles they take on few management functions, because of the potential contradictions between the objectives of a profession and the needs of a productive enterprise.

He proceeds to identify this as the major source of conflict that existed within surveying between contractors and private quantity surveying firms. As far as we are concerned here, the conflict may be described as not a contractor/private quantity surveyor conflict, but a large/small firm conflict. The issues raised at the time can be studied in terms of the reluctance to erode traditional professional values by an acceptance of business necessities. Similar issues can be identified, such as the abolition of fee scales to open up competition on price terms and lessen the importance of solely quality as a measure of output, and the various recent amendments to the rules of conduct, such as in advertising, the allowance of limited liability and membership, and the redefinition of what constitutes a chartered firm.

The archetypal traditional surveying firm is thus portrayed as a small partnership. Its organizational structure relies upon a paternalistic management system. Partners are powerful autocratic figures rewarding their employees' loyalty with pay and remuneration, working conditions, professional development and job tenure. As an alternative to a feudal structure, the organization can be compared with the family; partners are parents overseeing their childrens' progression to maturity and adulthood – which in this case is training for membership of the professional body and promotion through to eventual partnership.

4.3 THE RECESSION

Duality has now been seriously challenged. There are clearly emerging contrasts between the demands of the market and those of professional ethics. In periods of recession, it becomes increasingly clear that the simple aim of economic survival severely undermines notions of professional ethics. Response to change is increasingly pro-market and anti-professional orientated, resulting in radical changes to both firms and employees.

To examine this idea, consider the literature on the industry in Britain 6 years ago (September, 1989) with that now (September, 1995). In 1989, response to change meant largely dealing with a worrying shift in the demography of the UK (no school leavers, more women and elderly employees) and a view to the new markets of 1992 and the ever-unifying world economy. Now, change means that the firm must adapt to survive the recession, by consolidation, diversification and shrinkage.

While in 1989, the market was seen as the optimal resource allocator, in 1995 it is perceived increasingly as an anarchic distributor, in that firms can only react to its whims and are buffeted along randomly by the winds of market forces. Despite this, there is little movement towards state control, whereby centralized planning and predication could ameliorate these problems, due to the failure of state planning in Eastern Europe. There may be some tendency to look towards certain 'neutral' organizations to help, but these are not able to exert any control over the market, nor to exhibit constant and logical policies. Take, for example, the apparently idiosyncratic attitude of the RICS to key issues, such as its education policy and the Lay Report. How useful has RICS intervention been? Opinion seems to suggest that it has been of little assistance, but as it too is under the control of 'the Market', there is little it **can** do.

At a social level, employment is one of the most obvious changes between the 7 years. While it is obviously over-simplistic to suggest that in 1989 there were jobs, and in 1995 there are none, the recession does clearly illustrate a change in attitude by firms to their employees.

The primary change is that in 1989, employees were regarded as a firm's main asset – 'the human resource' – and the texts worked in the fields of motivation, payment by results and how to best keep staff. Today, they are laid off in order for the firm to survive, which throws the communitarian approach of the 1989 texts into a different light.

4.4 THE FLEXIBLE PROFESSIONAL FIRM

There is currently debate about the likelihood of small firm survival, either through the recession or some perceived take-over by the larger firms.

It is interesting to note that the ideals of the small firm are also put forward in favour of the larger firms, who are better able to afford training, provide better facilities, and generally look towards a better future for firm and employee.

There is thus a suggestion that large firms may prove the future of the surveying profession and form its organization archetype. Even if this is so, a more fundamental change has occurred as a result of the recessionary effects detailed above. Whether large or small, the professional firm has gone 'flexible'.

The flexible firm is a well-recognized structure in general management theory, but not to the traditional professional organizational form. The basic characteristic of the flexible firm is that employees are split into two types:

1. The core: elite staff are retained as full-time employees with excellent remuneration and fringe benefits. These are the essential professional staff and receive the treatment which well-motivated, well-qualified professional staff expect.
2. The periphery: the majority of staff are employed on some form of flexible basis: part-time, fixed-term or piece-rate are usual. As variable costs, these employees are paid low wages and expected to provide for their own continuing education, materials for carrying out work and welfare benefits.

While to many the working conditions in the periphery are familiar, to professionals, non-continuity of employment, low wages and no additional benefits are most certainly not. Indeed, flexible working conditions used to mean flexibility for employees who did not require, for whatever reason, full-time jobs, but still commanded high wages. not only has the periphery created a high skill–low wage labour market, but it also creates a high rate of labour turnover, aided by the large pool of reserve professional labour created by the recession.

The flexible firm is described as a lean, efficient organizational structure. It is a common business structure. It is, however, **not** a traditional professional firm structure.

4.4.1 The new professional (post-recession) firm

The current recession obviously demands action for survival by firms, and it is one of many modern forces that demand change. Surveying is, however, in a difficult position. Its professional ethics and working practices are based upon its Victorian heritage. The professionalism that surveying retains as its speciality is in increasing danger. This could prove catastrophic to surveying, surveyors and all those who rely on its professionalism.

This is a serious recession. Given the dominance of the market in both resource allocation and demand for services, there was little that firms could do except to react to the dictates of market forces. It is possible to consider a subjective model of this current situation and to recognize, in simple terms, a series of polarized styles of reaction. In a professional firm there may well be a communal attitude to the problem; everyone pulls together to survive the recession and act as theory suggests in a liberalized humane manner. The problem arises in the economist's typical business unit, where the professional managers have a different attitude to the problem. These managers act solely in their own short-term interest, and in a recession seek to cut costs to survive. However, when the recession ends, these firms may find themselves in a far worse position, as employees will remember those firms who offered permanence and security, and thus the higher order motivation needs of Maslow, Herzberg, McClelland *et al*. They could find themselves paying higher wages for less qualified staff, and suffering appropriately.

The post-recession firm firstly needs a post-recession! It is unclear at this point in time exactly how organizations will emerge from the recession, but it seems likely that firms will follow one of the following strategies:

1. Retain completely the flexible structure.
2. Revert completely to the traditional structure.
3. Adopt some form of compromise structure.

4.5 CONCLUSIONS

There are no definite conclusions. What has been proposed is a shift in the organizational structure of professional firms, coupled with a change in the attitude of individual professionals. The point at issue is the permanence of these changes. Have attitudes, especially those of employees, changed forever, or will they be forgotten with the next boom?

Much of the human resource literature has always been of doubtful use in its application to construction. Construction firms clearly fall into this flexible firm theory, where a small elite of core staff are retained, and the rest of the workforce is hired on a casual or part-time basis. Until recently, however, it had seemed that professional surveying firms adapted a rather more permanent structure; professional staff, according to these same motivations theories, requiring a rather more liberalized attitude. The current situation has thrown this into doubt.

Traditional theories have been severely tested, and it is doubtful if they can recover. This rather disturbing picture has shattered the pre-recessional view of a professionally devised response to pure market forces, and a belief that the surveying profession can manipulate those forces. It is tempting to romanticize the pre-recessional control that the professional institutions wielded in maintaining quality, but it should be recognized. The arguments were rehearsed in the debate over standard fees as a guarantee for competition purely on quality terms, and lost at that time. While professional integrity survived as a major issue before the recession, mere business survival and job retention are in danger of having terminally undermined the previous ideals, despite the threat of professional monitoring and increased expulsions for malpractice.

In a market economy of price competition, this is not surprising. Surveying as a profession has been fortunate to have survived for so long relatively untouched by such forces; general industry has long been under its power, and the majority of the construction and property industries have for some time looked at the nonchalance of surveying with awe. Notions of professionalism cannot, in the long run, survive in a market economy.

4.6 REFERENCES AND FURTHER READING

Atkinson, J. (1984) Manpower strategies for flexible organizations. *Personnel Management*, August, pp. 28–31

Avis, M.R and Gibson, V.A. (1987) *The Management of General Practice Surveying Firms*, University of Reading.

Ball, M. (1988) *Rebuilding Construction*, Routledge.

Barrett, P. and Males, A. (1991) in *Practice Management* (eds P. Barrett and A. Males), Chapman & Hall, London, pp. 3–12.

Best, M. (1990) *The New Competition*, Polity.

Braverman, H. (1974) *Labor and Monopoly Capital*, Monthly Review Press.

Cleary, M. (1992) European Standard. *Chartered Quantity Surveyor* (May), **14**(9), 11.

CSM (1992) Chartered designation – questions of supreme importance. *Chartered Surveyor Monthly* (September), **2**(1), 5.

CSM (1993) Members vote for diversification. *Chartered Surveyor Monthly* (January), **2**(4), 4.

CQS Editorial (1992a) *Chartered Quantity Surveyor* (November), **15**(3), 5.

CQS (1992b) RICS says 'yes' to 75% rule on chartered designations. *Chartered Quantity Surveyor* (November), **15**(3), 21.

Davis, L. (1992a) Bucknall's bright idea. *Chartered Quantity Surveyor* (November), **15**(3), 20.

Davis, L. (1992b) Three into one. *Chartered Quantity Surveyor* (October), 15(2), 20.

Eade, C. (1992) Chesterton Chairman slams RICS action on Lay report, *Chartered Surveyor Weekly*, 13 February, p. 7.

Eccles, T. (1993) A Philosophical Review of Professionalism in Surveying. Occasional Paper No. 2, Kingston University.

Eccles, T. (1995) The Quantity Surveying Firm As A Unique Business Unit. Occasional Paper No. 8, Kingston University.

Estate Times (1992) Wells raps RICS over Lay report response. *Estates Times*, 14 February, pp. 1–2.

Lay Committee (1991) *Market Requirements of the Profession*, RICS.

Mallett, L. (1992) Change or Die. *Estates Times*, 14 February, p. 9.

Poole, R. (1991) *Morality and Modernity*, Routledge.

RICS Junior Organization (undated) *RICS Fit for the 21st Century*, RICS.

Schumacher, E. F. (1973) *Small is Beautiful*, Blond & Briggs.

Sloan, B. (1991) Recruitment in the 90s and beyond – towards an understanding of the effects of the demographic changes on the construction professions, in *Practice Management* (eds P. Barrett and A. Males), Chapman & Hall, London, pp. 35–40.

Thompson, F.M.L. (1968) *Chartered Surveyors: The Growth of a Profession*, Routledge & Kegan Paul.

Torrington, D. and Hall, L. (1991) *Personnel Management: A New Approach*, Prentice-Hall.

Turner, A. (1993) Professional Prostitution: open letter to the president. *Chartered Quantity Surveyor* (December/January), **15**(4/5), 7.

Watts, T. (1992) Presidential Address 1992. *Chartered Surveyor Monthly* (April), **1**(7), 8–9.

PART TWO

Managing Professionalism and Creativity

Introduction to Part Two

Sir Michael Latham, Chairman of the Construction Industry Board

It is a particular pleasure to introduce this section of the book. Managing professionalism and creativity is an intriguing theme, and I shall seek to develop in a moment. It is, if I may say, absolutely within the wide-ranging approach which we have come to expect from those institutions that are still called new universities – but I prefer to call universities. They are challenging the old and traditional authorities in many ways, including in the delivery of new courses and fields of study. I do not know how long they will still be referred to as new universities, but I suppose we still have the New Testament after some 2000 years! I am also grateful to Liverpool John Moores University, as part of the International Procurement Review Group, for the evidence which it submitted to the review of the construction industry which bears my name, and which is quoted in the final report. It draws attention to the need for more interdisciplinary work at undergraduate level, but not at the expense of specific skills and competencies.

It is also very appropriate that the conference on which this book is based took place at the Institution of Civil Engineers, of which I am privileged to be an Honorary Fellow, though my engineering skills are, to put it mildly, modest, not to say non-existent. We held the second of our conferences for the 12 Working Groups which report to the Construction Industry Board there. Of these 12, one has finished its work, another is engaged upon information of its findings, and several of the remainder are coming to the end of their tasks. Our purpose was to seek to ensure that the outputs of the Groups are consistent with each other, and also with the general aims of the CIB. This is no easy task, because those aims are disparate. Some of them are directly relevant to the theme of this book, and all of them affect it to a greater or lesser degree.

I would now like to identify several themes which I believe are encompassed by the title of this section. But first, a word about the title. I suppose it is relatively recently that anyone would have thought that such a title was appropriate or meaningful at all. To **manage** professionalism?

What could that imply? Surely professionalism has a description of its own, identified by its codes of behaviour and ethics, routed in education and a great tradition of integrity and duty? Why did it need to be **managed**? Was not its role to manage others, especially in construction? And what could **managing creativity** possibly mean? To a creative person – especially perhaps to a designer – his or her skill is partly innate and partly acquired. Their artistic talent was capable of development and nourishment, of course, but management is the least of the requirements for developing it. Did Beethoven or Raphael need management skills? Did Michelangelo? Among the great architects and engineers, a pair such as Wren and Brunel, had they ever had training in **management**? Such thoughts would certainly have gone through the heads of people receiving details of this book some years ago. Indeed, I wonder if such a book could have been successfully published at all.

I think it is a mark of how things have changed in construction that such expression as 'managing professionalism and creativity' no longer seems quaint, or simply philistine. And the main reason for this is the emergence of the professional client, at the centre of the process, who is looking for a construction team which will achieve his or her objectives, rather than one which will tell them what they ought to have. But not all clients are professional. Many are lazy and unknowledgeable about the construction process, including some of the newer processes in the public sector. This offers opportunities, but also challenges, to the professional and calls on their experience. I include of course professional constructors and specialist contractors as well as consultants. The client needs help and guidance to devise an appropriate procurement strategy. This is an exercise which proceeds briefly, as I tried to suggest in my report. There are several questions which a client considering a construction project should be asked: Do you actually need a project at all? If so, what is your policy of risk acceptance or transfer? How does that risk decision affect the potential choice of procurement route? Having chosen an appropriate route to suit your project strategy, what steps should you take to retain people able to devise your contract strategy, including formulation of the brief by an intensive process and assembling the right team? These are all, in a real sense, management decisions. Some might say **project** management decisions. I prefer to see them at this stage as putting the project strategy on a sound and business-like basis. Such professional advisers may then go on to a wider role to become lead managers, lead designers, project or construction managers, or under more traditional but equally purposeful descriptions as architect or engineer. But the initial role could equally be the whole of the commission, and such a limited but dignified role in independent project assessment is envisaged both by the Central Unit or Procurement and by using an engineer.

Every project needs to be managed. I have never really been sympathetic with the argument that the basic needs of the client change when the design begins to manifest itself in reality on site. It does not really matter what the discipline is of the person responsible for managing a project, in that there is no specific discipline, in my view, which is inherently likely to be better at management than others. Some modern architectural practices recognize this very clearly by differentiating between those whose skill lies in creative design and others in contract administration. They can, of course, be combined well by one person and doubtless are on many occasions. But they are not the same skills. The client may wish to leave supervision to a professional consultant, but delivery of the project to the contractor. But they may also prefer to have one person clearly in charge of delivering the entire project, ranging from advising on the initial strategy, hiring all the downstream consultants, ensuring the integration of design and construction, and in particular the initial role of specialist subcontractors with their heavy input of detailed design, and then acting as the mover and shaker from start to finish. That role may be in-house, may be retained as a consultancy, it may be seconded to the client, or it may be transferred to the contractor under a design/build arrangement, by which risk is transferred as well. But it must involve at every stage a high level of management skill, and one which cannot simply be taught. It is right and proper that universities are increasingly offering such courses as a normal extension of the curriculum, since they recognize that a new discipline has emerged – and is not going to disappear. The very role of project manager remains controversial, in that some commercial clients still see such a professional as little more than a glorified postman, without an additional level of fee, while at least one large developer described having to make such an appointment to me as a defeat. But as the rules and duties become better defined, it can be seen that there is a real task here. It is also right, in my personal view, that most universities see it as a postgraduate one, to be pursued by the experienced practitioner who already has significant experience of conditions, and of the practical problems of the industry. I am not, of course, saying that project management as a discipline, with possibly interchangeable skills of management, cannot be taught at any level of age and experience. Doubtless it can, but more hands-on experience may make for a more effective approach in construction.

There is certainly still scope for more training and a clearer status for project management within the industry and a need for a single and generally agreed list of duties and responsibilities for project managers. I recommended this in my report, and it is being pursued in working groups. In a sense it is like trying to keep up with computer technology.

As far as one feels, the duties and role are clearly defined, they emerge or evolve further, as new procurement techniques or refinements of them are tried by progressive clients.

Which brings me to a related and difficult question for a modern professional, which is **positioning**. There are a number of reasons why I have been supportive of the New Engineering Contract, or alternatively of seeking to restructure the standard conditions to meet what I believed to be necessary principles of a modern contract. They include, of course, the emphasis on teamwork and win-win solutions, and rewarding good management practice. But they also include a very clear division of responsibilities. The main parties to the contract are the employer, as represented by the Project Manager, and the Construction Engineer, their specific title for the moment, as it could equally well be lead manager or client's representative, as in the BPF form. The Project Manager, under that designation, can undertake other duties under the NEC, including the role of designer and supervisor if they so wish, though these functions can also be represented. But the role that they cannot perform is to be the adjudicator. The person who is clearly positioned in the client's company cannot also be the referee. That is one clear role of the NEC. The other is that there is no place for nomination.

Now it may be argued that architects and engineers are specifically instructed during their professional education to act as impartial arbiters between client and contractor when disputes arise during contract administration. Indeed they are, and Contract Conditions such as JCT 80 or ICE 5th and 6th specifically reflect that. The problem lies not with the integrity of the professionals, but with the scepticism or reluctance of the client as their employer, or the contractors as the other party to the contract, to accept a decision which may reflect a mistake by the designer or alleged slow delivery of documentation. There are also commercial pressures upon the professional which I do not need to spell out for you. I believe that it is essential for the efficient management of a project that the lead manager, call him what you will, should clearly be seen to be of the clients group – something, incidentally, that clients also want and cannot understand if they do not get since they assumed that that is what they are paying for. I believe that adjudication by a nominal third party should become the normal method of dispute resolution in construction contracts, and be underpinned by legislation. Adjudicators will, of course, require to be trained and governed by a Code of Conduct. However, this is not a new discipline, since procedures already exist in several contractual forms for adjudication. Nor do I believe that it will be a particularly remunerative one. Since the basis of underpinning by contract and legislation will be to require the adjudicator's award to be implemented immediately, even if subsequently appealed to arbitration post partial completion, I think that the likelihood is that most disputes will be settled between client and contractor, or contractor and sub-contractor without course to an adjudicator. Such experience as exists of contracts with these procedures confirms that their use is the exception rather than the rule. And if contracts can be amended to lay emphasis on research, joint ratification and discussion of problems as they are seen to be arising, and resolution of them jointly, the need for an external decision maker should be correspondingly less. It also significantly reduces the burden on the lead manager or contractor administrator, especially if he or she is also the designer and supervisor as they can then concentrate upon their client-supporting role. But not only this, they can also concentrate upon their creative and managerial functions fused together, in seeing the scheme as built, first on their merits and then helping to bring it to practical fruition.

Another aspect of growing importance to creative managers is empowering the client at the earliest conceptual stage. The technique of knowledge-based engineering, or virtual reality videos, is expanding rapidly. When I began my review, it had been discussed by a few experts. As the months proceeded, I heard more about it and went to look at work being done at Reading University, on behalf of BAA plc. I found this such an exciting concept, and so full of possibilities for assisting lay clients to understand their projects at the earliest stage, rather than grappling with architects' drawings which are not easy for a client to grasp, that I asked Reading to put a paper to an Assessors Meeting in early 1994. When the paper arrived, it was greeted with polite scepticism by the contractors, who saw it as the technology of the 21st century. They thought that the software would be too expensive, that clients would not be prepared to pay for it, and that architects and engineers would not be able to afford it because of the meagre fees which they would obtain through competitive fee bidding. But the client assessors thought differently. They saw software costs falling quite rapidly, and clients increasingly being prepared to pay for a management tool which could prevent the disaster scenario with which we are all familiar, whereby the lay client comes onto the site half-way through construction and says, 'Oh, I didn't realize there was going to be a window joint there – please move it, it isn't right there'. We all know the response, and it can be summed up in the immortal words of the radio quiz game, 'I'm sorry, I haven't a clue'. Those words are '. . . it's going to cost you' – and it does!

We now know that the client's hunch has been proved correct. Software costs are falling and the technique is being increasingly used by clients, or demanded by them. This empowerment is of the highest possible importance for all involved in the construction process, not least because it prevents misunderstandings before they arise, and can itself be an important driver down of unnecessary and wasteful costs which add no value to the project. These techniques should become a standard part of the professional education of designers, since they combine both management and creativity. It is an exhilarating experience to see the video of BAA's proposed fifth terminal at Heathrow. But South Glamorgan County Council use similar techniques in their schools' architecture department, and no doubt other clients are increasingly doing the same. It speaks volumes for our industry that while such new techniques are initially seen as fanciful and for the 21st century – which is only 5 years away in any case – we are still far from having coordinated project information adopted as a normal practice, let alone a contractual requirement. When the assessors finished their discussion on virtual reality, they went on to talk about CPI. One of the client assessors commented dryly that if virtual reality/knowledge-based engineering was to be regarded as the technique of the 21st century, CPI should be seen as that of the 19th century. 'Surely', the client said, 'we can harmonize the basic works information?' Again, clients must be prepared to pay for a proper service, because they will ultimately be the losers if they skimp on the up-front costs, as best-practice clients understand very well.

And that brings me to the issue of remuneration for creativity. Speaking personally, I do not like compulsory competitive fee bidding as a route for selecting consultants. I agree with a very large and experienced private sector retail client, with an annual spend of umpteen millions, who told me that he would never dream of selecting a consultant on such a basis, and always used negotiation. He did not pay what the contractor asked for, still less contemplate a scale fee, but he did not believe in a sacrificial fee either. He wanted the best service, and expected to pay for it. If he did not get it, he looked elsewhere next time.

However, that is not consumer policy, and public authorities are required to use a competitive fee bidding route. Given that policy, which seems unlikely to change, it is all the more important that consultants should be chosen on a basis which reflects quality as well as price. It is not difficult to do this, and indeed several models were already in existence or about to appear when I published *Constructing the Team* in July 1994. The need is to give political comfort to public sector officials who feel very threatened by ambit or other pressures if they do not choose the lower bid, particularly if they have been through a responsible pre-qualification procedure first. They know that, technically, they do not have to choose the lower bid, because guidance from the European Union Treasury Central Unit on procurement and the National Audit Office here all stressed that they should choose the best or most advantageous bid, not necessarily the lowest. They know also the questions in the White Paper *Setting New Standards*, which lays very clear emphasis on quality and similar factors other the price. But they still feel the need to be able to point to a semi-mechanistic route, and be able to say to auditors or other objectors that they followed the quality/price mechanism set out in an official document, and that these were the criteria which led them to choose Smith, Rocket and Jones rather than Brown, Williams and McDonald. Our Working Group – whose chairman incidentally is a major private sector client, Jeffrey Wright of Harrisons – is very shortly to produce for the CIB its quality–price relationship report, hopefully for the Board's meeting in October. If approved, it will then become available for use, and the sooner the better. When I compiled my report, I was extremely concerned about the strong feeling, almost a sense of despair, among professionals, and especially architects and engineers, about the rigours of fee competition. A survey conducted by the ACE was particularly depressing. Although the response of some cynics was a hoarse laugh and '. . . well they would say that, wouldn't they', others found some of the findings very disturbing. I was neither particularly surprised nor dismayed that members of the ACE were prepared to tell their own organization that 94% of them bid low to maintain cash flow or test the market, or that 79% were spending less resources on training graduates and technicians. But what was one to make of the 74% who admitted that they were producing simpler designs to minimize the commitment of resources to a task, the 61% who bid low with the intention of marking up fees with claims for variations, or the horrifying 12% who were prepared to tell a trade association to which they belonged, in these days of CDM, that they paid less attention to health and safety, both in design and on the site? That is not the kind of admission anyone would normally dream of making, and that alone, if nothing else, required further thoughts of how we could help public sector clients in particular, to balance quality and price in their assessment of competitive fee bids.

When my report first appeared, at lease one distinguished architect gave his initial reaction that the report did not say enough about 'why', concentrating instead upon 'how'. He also felt there should have been more about design quality and the wider environment. Those criticisms may well be fair, though in fact I believe that he softened them later when he read the report more carefully, tried to assess its overall balance and nuances, and compared the initial feelings which he had with the meetings he had with me and the speeches which he heard me make. To some extent, my findings were governed by my terms of reference, which were forwarded to me by those who commissioned the survey and which only once used the words 'design' and 'quality', and in both cases as part of a wider issue. Thus, I did my work with regard to responsibility for the production, management and development of design, and took into account, among other things, investment in improving quality and efficiency. Nevertheless, I forced into the report an emphasis on quality and cost-in-use, and on the role of the client in patronage, including the public sector, despite the unpromising background against which I had to compile my recommendations. You may be amused to know that I have three times been told by experts that they had believed I was an architect, the most recent occasion being an academic in a University Department of the Built Environment. Well, I am not an architect, and neither am I a lawyer, of which I have also been three times described. But quality and cost-in-use must come much higher up our list of priorities that they have in the past. They are more likely to be achieved if proper attention is given to stripping out of the cost equation those elements which add no value, but simply reflect hassle factors. Our Working Group is making very good progress in that regard.

The construction process is one which affects every-body all the time. We cannot all see great masterpieces of the visual arts such as the works of Raphael or Rembrandt. For that we have to go to galleries or great stately homes. But we can see the results of the creativity of our designers all around us every day, whether in buildings, civil engineering or process engineering projects. We cannot calculate genius or creativity, though we can nurture, sustain and influence it. What we can, and must do, is to ensure that it is able to develop in an environment which we use it to the best advantage, a client-centred environment which is also conscious of its wider responsibilities and duties. Such an environment will be more receptive if it knows that creative skill is also harnessed to an efficient realization of a modern construction process, which is cost-efficient, avoiding waste or excess, and realising the objectives of the client-pattern. That is what this section is about. I am sure it will be developed in the chapters which follow.

CHAPTER 5

Developing creativity as a core skill

5.1 INTRODUCTION

Creativity, creative thinking and creative problem-solving techniques have developed rapidly since the 1930s and 1940s as a philosophical discipline with practical applications to the changes facing the management of all sectors of industry and commerce. As such, it has the potential to enhance the strategic and operational performance of the land, property and construction industry and its professionals.

5.2 CREATIVITY

Creativity has been variously defined by many practitioners over the years, principally from a philosophical perspective. Welsch evaluated the degrees of agreement between the various, sometimes conflicting, definitions and has defined creativity as:

> the process of generating unique products by transformation of existing products. These products, tangible and intangible, must be unique only to the creator, and must meet the criteria of purpose and value established by the creator.

. . . a definition highly suited to the property and construction industries.

Other key indicators of creativity and creative thinking include reference to aspects such as:

- unique and interesting associations
- the making and communicating of new connections
- the thinking of possibilities
- the consideration of new points of view
- innovation
- originality of thought

All these component parts fit well with the processes of design, production and management of buildings and property.

The degree of creativity that can develop in an organization depends on the level of freedom to generate:

- an intuitive creative atmosphere;
- a non-controlled creative atmosphere; and
- a formalized creative atmosphere.

In each case, the common creative processes take place in the mind of the individual and tend to range between convergency and divergency.

5.2.1 An intuitive, creative atmosphere

This is one that possibly in combination with an element of luck, results in a state of serendipity (an almost casual stumbling across something that is interesting). Classic examples of this include the landing of the apple on Isaac Newton's head and the concept of gravity and the cry 'Eureka!' as Archimedes displaced water in the bath.

5.2.2 Creativity within an inspiring, non-controlled atmosphere

Examples here may in effect be market- or customer-driven, provided that the culture or setting allows for exploration of the 'new', irrespective of whether it relates to products, services, marketing strategies or any other operational or strategic activities. In such a facilitating atmosphere, the balance of positive opportunity and negative threat is not an issue. The individual is allowed the freedom to create! An architect conceptualizing in a broad-based practice, a valuer and a land surveyor operating within a small team on site have the potential to demonstrate such freedom of creativity. An architect in many respects has his or her professional integrity measured by performance indicators that in effect quantify creative output.

5.2.3 Formalized creativity

This is often the main focus in many professional practices with regard to issues such as the analysis of customer problems and needs, specified operational activities and strategic objectives by processes such as brainstorming and team meetings. They can be the most cost-effective when measured as direct, performance indicators, but the process of focusing can result in a degree of blinkered, constrained, even inward-looking attention that does not allow for consideration of the 'big picture'. An architectural technician may well fit into this categorization when the practice has a limited range of client types and the technician works on a restricted part of larger projects. Quantity surveying and estimating, facilities management and services engineering also tend to operate within these constraints, their ability to create being somewhat restricted by the limited focus of activity.

In the context of the built environment, the philosophy of creativity thus allows for direct impact on issues ranging from the original concept of a new building via the consideration of new approaches to site-related logistics through to the practices employed in estates and facilities management.

This is more clearly apparent when taken in the context that every building or structure is unique in terms of its size, shape, location, context and client–contractor relationship and hence has the potential to allow for unique solutions, procedures and practices.

This does not imply that each new contract starts with a clean sheet of paper. The emphasis should be on creative learning from past and current experiences, to establish new and different approaches to solving and managing current and future problems.

5.3 RATIONALITY

Rationality is part of the creative process whereby the individual, the professional practice or the organization, analyses and evaluates the reasoning and the implications for the practice in the longer and shorter term.

Creative issues need to be freely addressed such as:

- 'What is our business?'
- 'What business are we in?'

The two answers may well be different! And for example:

- 'Are we house builders?' or
- 'Are we a business that creates jobs in the housing industry?' or
- 'Are we a business that aims to make a profit with house building purely being the nature of our output?'

This rationality in turn can lead to action, critical thinking and maximized creativity. All team members, individuals and practices need to be committed to achieving the 'best', often expressed as quality or excellence, or in creating in a better way, a level of creativity that is more skill-based, more focused, more thoughtful and more careful.

Rationality has characteristics appropriate to a unique culture, time or discipline. For example, it is probably impossible for surveyors, architects, planners or builders in the 1990s in the UK, to fully rationalize the building of the pyramids or Stonehenge, despite our high levels of technical knowledge, our collective experiences, our ability to experiment and to test models and our enhanced understanding of the physical sciences.

Rationality is thus based upon many issues that include:

- a clarity of analytical judgement, supported by all relevant evidence, sometimes referred to as a 'total evidence condition';
- use of logical criteria, such as evidence and belief with some inductive basis;
- use of informal logic, to assess the adequacy of the various argument forms to provide justifiable 'good reason';
- probability theory; and
- degree of knowledge and understanding.

But this rationality can be a liberating or a dominating tool depending on how it is used or misused. It thus becomes essential to differentiate between genuine rationality and 'apparent rationality', which leads to mistaken action.

Classic examples of misuse are in attitudes and actions related to racial, social, ethnic, cultural and sexual discrimination. Apparent rationality for example is inherent in the social and ethnic arguments put forward by states that implement 'ethnic cleansing' actions.

A degree of rationality is associated with every planning application, every example of property design and every bill of quantities. Yet some proposals are accepted as others are rejected, while some are accepted in the current context only to be built, criticised and subsequently demolished.

Because of this cultural context, there may be a need for trans-cultural application to facilitate the transferability of learning, both by the individual, the professional and the technician and by the practice, client and contractor. To convince an individual that their rationality is not genuine requires:

- an openness to discussion;
- an awareness that judgement may be wrong;
- an ability to enable the individual to be comfortable in accepting change; and
- a range of techniques to facilitate the change.

Yet these needs are in themselves, examples of the key components of creativity. Creativity is thus a function of rationality, but rationality is also a consequence of creativity.

5.4 A CREATIVE CYCLE

A systematic approach to creativity needs to refer to a series of stages as part of a learning cycle – the 'creative cycle' – in which rationality is only one of the associated issues (Figure 5.1).

Although philosophers traditionally tended to consider the concept of

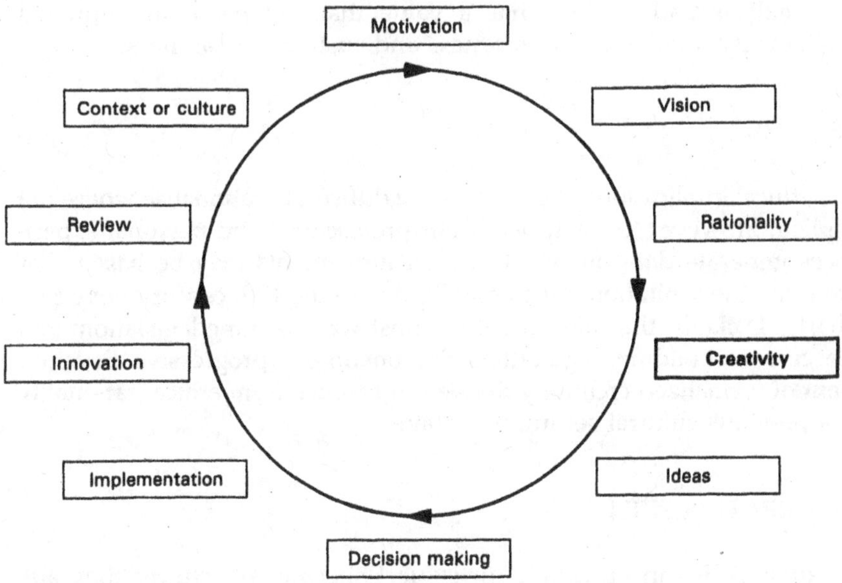

Fig. 5.1 A vision of creativity – the Creative Cycle.

creativity in an idealized form, a number of strategic thinkers have taken this a stage further in its application. A creative vision for an organization or practice is seen as one that is articulated clearly, forcefully and consistently.

Recent management trends in the application and understanding of Business Process Re-engineering (BPR) is in effect a creative movement towards a vision of radical corporate change that requires the organization in its entirety to restructure and recreate itself at a level that is efficient and effective. Tom Peters (1987) for instance has proposed that

> Managers must create new worlds. And then destroy them; and then create anew. Such brave acts of creation must begin with a vision that not only inspires, ennobles, empowers and challenges, but at the same time provokes confidence enough, . . . to encourage people to take the day to day risks involved in testing and adapting and extending the vision.

Similarly, Majaro (1991) stated:

> Before one can improve the firm's performance in this regard (to improve creativity) one must attempt to audit the level of creativity that exists in the organization at a given point in time.

In simple terms, creativity as a resource can be perceived as an aid or systematic process to facilitate problem solving, decision making and risk analysis and as such has a value that can result in improved productivity, market responsiveness and customer awareness.

5.5 ETHICS

The ethical implications of creativity are difficult to rationalise where not applied. However, once applied, our professional and personal experiences generate data on which ethical judgements may be based. For example, the sanitation issues highlighted in the 19th century novels by Charles Dickens, the subsequent UK post-war planning legislation, and the current building regulations demonstrate a progressive development of formalized creativity derived in essence from ethical rationality in a previous cultural setting in Britain.

5.6 CREATIVE SKILLS

Creative skills can be generic or subject specific. Whichever they are, they need to involve aspects such as consequentiality, prioritization, contrast, assessment of options and final choice. The key skill though is to be flexible. Cross (1989) refers to creativity in problem solving as being a life-jacket and not a straight-jacket, with the skills and techniques being applied to the individual's own thinking style.

Techniques to facilitate and develop creativity should start early in formal educational development, as the common-sense, informal educational process of learning ('The University of Life') does not fully develop the skills in all individuals.

University training for built environment professionals, the continuing professional development programmes and skills updating through short courses and flexible learning, allow the individual and the professional practice to develop and enhance creative thought processes and creative practice. Techniques such as role-play, brainstorming, rehearsal and others can be enhanced and developed, the actual choice being dependent on the degree of creative freedom within the practice.

Practices that operate an informal and intuitive creative process can make better use of techniques such as brainstorming, brainwriting, checklists through to random stimuli and analogies. More systematic and formalized practices may gain benefits from listing of attributes and use of morphological techniques:

1. Brainstorming can be used by small groups to generate many divergent solutions to a problem with no restriction on the initial appropriateness of any proposal.
2. Brainwriting provides a similar option for the individual or the team. It may provide the stage before brainstorming.

3. A shopping list of controlling or restricting issues may additionally be used as a trigger to determine potential solutions. For example, time penalties . . . plus availability of plant . . . may result in implications for multi-skill training.
4. Random words may also be used to generate apparently unconnected links, e.g. (i) ship . . . floating . . . applications to concrete?; (ii) ball point pen . . . ink flow . . . applications to concrete flow control?
5. Attribute modification whereby existing solutions are applied to new problems albeit in a modified way. For example, do the materials used in Formula 1 racing cars have application to fire restriction in domestic construction?
6. Morphological analysis according to function provides for different combinations of ways to achieve the same quality of solution. It is in effect a localized version of Business Process Re-engineering.
7. Role-play can help for example by putting yourself in the historical role of the chief architect or planner to the ancient Egyptians or the medieval castle builders to enable the individual to assess what he or she would have done in the same circumstances and what degree of creativity could have been employed.
8. Rehearsal of the progression through the stages of a project, allows for value-based judgement of consequences before passing the point of no return during the construction phase. It allows for a more deliberative analysis and assessment of risk through standing back, evaluating alternative strategies, and consequently lessening the risk.

Training and learning for students in the built environment needs initially to allow for the application of creativity in a safe environment, evaluating typical projects and later more complex tasks.

The skill of creativity to some extent is a natural process but this can be superficial in a world of rapid change. Formal built environment education needs to challenge this and prepare for it by making creativity an explicit component of built environment learning, which ideally builds on appropriate learning and understanding from the school.

5.7 DEVELOPMENT OF CREATIVITY IN THE PROFESSIONAL PRACTICE

Built environment education for the professional practice needs to integrate a number of creative dimensions

1. Motivation, for example, needs to address the issues associated with the individual who puts forward a proposal that is not consistent with practice policy or the individual who does not feel free to state such proposals.
2. Team management need to address the issues concerned with the

individuals who currently work together and the future needs of the practice. New recruits may be required who will bring in new creative ideas and be allowed to fully use them within a creative atmosphere.

3. New managers who facilitate rather than block the development of new ideas may be required.
4. All individuals should be allowed the capacity to create, even if each individual expresses it differently. It is equally valid and equally important in terms of quality for an architect (for whom creativity is in many respects the norm) to utilize creative thinking in the design of a building as it is for the groundworker who expresses his opinion of the architect's ability when he is faced with a site problem resulting from a design error. Creativity is fully enhanced when the groundworker's views are respected and acted upon.
5. The process of construction, either *in situ* or with prefabrication, is a dynamic process that should allow for reconceptualization based upon learning experiences, historical trends and scientific enhancement.
6. Individuals should be allowed the freedom to experience other activities or processes in the system which as a result will enable greater creativity to emerge. For example, does the buyer carry out his or her activities as a function of a particular contracts needs, or as a function of the longer-term needs of the contractor?
7. In evaluating the reasons for excess project delay, other peoples' views – including the sub-contractors – may be beneficial in the longer term!
8. Individuals and teams must have the motivation, means and opportunity to be creative. All are essential.
9. A professional is accountable for his or her actions. Creativity is a component of accountability.

5.8 REFERENCES AND FURTHER READING

Cross, N. (1989) *Engineering Design Methods*, John Wiley, Chichester.

Isaksen, S.G. *et al.* (1993) *Understanding and Recognising Creativity; The Emergence of a Discipline*, Ablex Publishing Co., Buffalo, USA.

Isaksen, S.G. *et al.* (1993) *Nurturing and Developing Creativity; The Emergence of a Discipline*, Ablex Publishing Co., Buffalo, USA.

Majaro, S. (1991) *The Creative Marketer*, Butterworth-Heinemann, Oxford.

Peters, T. (1987) *Thriving on Chaos*, Harper Collins, New York.

Sharkey, M. (1995) Management of Creativity and Developing a Creative Culture (for the Built Environment). Second International Conference on Creative Thinking, University of Malta.

Welsch, P.K. (1980) *The Nurturance of Creative Behaviour in Educational Environments: A Comprehensive Curriculum Approach*, unpublished PhD Thesis, University of Michigan.

CHAPTER 6

Human resource management and structured training

6.1 INTRODUCTION TO HUMAN RESOURCE MANAGEMENT

This chapter is concerned with the changing nature of personnel or human resource management (HRM) activities in professional practices and the challenges facing those responsible for such activities, especially for the structured training of professional staff, which will be considered later. To provide a framework for understanding some of these changes, a model of HRM activities in professional practices is presented. The model emphasizes the role of HRM as an agent of change and outlines some of the skills required by practice managers responsible for such activities.

Professional practices – particularly in the fields of land, property and construction – are often referred to by professionals working in these fields as being primarily 'people businesses'. 'Our only assets are our people' is another view which is also regularly expressed by senior surveying professionals. Because of the vital importance of people in professional practices, investment to foster the growth and development of these professional resources is a major factor in achieving business success.

Ensuring that practice managers design 'people' policies and implement procedures which will have maximum benefit is a responsibility which is, in many instances, not explicitly defined and one which is often shared. In small to medium-sized practices such responsibility is likely to reside primarily with the Senior Partner, while in larger and more geographically distributed practices it is more likely to be shared with other Partners or Directors, alternatively it may be devolved to the 'Staff' or 'Managing Partner'. In the largest practices, specialist expertise may also be provided by personnel/human resource (HR) staff.

This section is concerned with some of the roles which need to be performed by those senior professionals who have ultimate responsibility for developing the 'people policies' of a property-related service sector business. Before describing some of the particular characteristics of these aspects of HRM, it is appropriate to consider some of the specific features of professional service organizations and the consequences which such features have on those who provide an internal service such as personnel/HRM (section 6.2). The paper subsequently examines the growth of human resource management (HRM) as a concept (section 6.3) and the development of the personnel function within business. A model is then proposed which links HR activities more directly to business objectives (section 6.4). The paper concludes by considering the skills required to adopt a more strategic approach to complement and supplement the more frequent administrative activities traditionally associated with 'staff' or personnel management.

6.2 SOME CHARACTERISTICS OF PROFESSIONALS AND PROFESSIONAL SERVICE ORGANIZATIONS

In order to fully appreciate the particular nuances of personnel/HRM in professional service organizations (PSO) and the specific challenges associated with its practice, it is helpful to understand some of the features of both professionals and PSO. The nature of professionalism is subject to a variety of different interpretations, which will be examined in greater depth in subsequent chapters. Suffice it to say here that the general attributes of professionals tend to be:

- higher education qualifications (which illustrate an ability to learn and to amass new knowledge);
- intellectual skills and articulacy (showing an ability to grasp new events quickly and to respond effectively and creatively);
- high levels of individual autonomy and significant degrees of discretion in the workplace (showing an ability to assume multiple responsibilities and self-management when discharging them);
- intrinsic motivation (illustrated by a strong desire to achieve); and a strong loyalty to the professional discipline and individual clients (which is often much stronger than loyalty to the organization) and which may act against cross-functional team working.

If one examines the nature of professional work, it is also possible to identify the consequences which these characteristics have on the approach which is adopted when practising HRM. Firstly, because the products supplied by professional service organizations are generally non-standard, as mentioned above, professional discretion is paramount. Appreciating the need for a degree of discretion in practice management matters is also vital. This must also be real and not contrived if one is to avoid damaging consequences. Secondly, the delivery of professional services is generally very highly time-constrained, and professionals have a low tolerance for administration and 'bureaucracy'. Consequently, to a much greater extent than in other types of organizations, those responsible for HR matters need to ensure that their activities avoid professionals expending time on unnecessary administrative matters, while at the same time ensuring professionals value the importance of people management and development.

Thirdly, because of the higher degree of professional discretion which exists and the flat organizational structures, reporting lines, the setting of performance criteria the review of performance is much more complex than in the case of more traditional, hierarchical organizations. Fourthly, professionals generally perceive 'management' to be a relatively low-status activity; consequently considerable explanation and persuasion is required in order to demonstrate the added value which can accrue from such activities. In an increasingly cost-conscious and competitive environment where effective management can be the critical factor in achieving success, there is clearly a potential tension between the practice needs and the beliefs of individual professionals towards management. Indeed, many would argue that the key role of management should be to service and support professionals in their dealings with clients. On this basis professionals may then be willing to cede authority to those in practice management roles to enable them to manage the organization. Above all, HR specialists in such environments will be expected to deliver high levels of service to their internal clients, mirroring the service level expectations which clients demand from professionals. The internal client relationship is also critical in relation to other service functions. In addition, and in a similar manner to the role of the professional, it is vital to build effective internal relationships on the basis of trust and referrals from others within the organization. Working with and through networks is also important if success is to be achieved.

A more detailed analysis of the changes affecting professional and PSOs and the leadership and management models appropriate in such cases can be found in Middlehurst and Kennie (1995).

6.3 THE DEVELOPMENT OF HUMAN RESOURCE MANAGEMENT (HRM)

6.3.1 Definitions

The concept of HRM is a relatively recent phenomenon in the UK. In common with many management changes its origins can be traced back to the United States. It grew in emphasis primarily through the 1980s leading to publications which emphasized some of the 'apparent' differences between the traditional personnel role and the new HRM role. The following quotations give some indication of the range of views which have been expressed about the matter.

> HRM involves matching HR activities and policies to some explicit business strategy – to align the formal structure of the HR systems so that they drive the strategic objectives of the organization. HRM is seeing people as a strategic resource for achieving competitive advantage. (Henry and Pettigrew, 1986)

> HRM is simply a retitling of the Personnel Department . . . its interest lies more in what it says about the poor image of personnel management than in anything new that it offers. (Guest, 1989)

A third view of HRM which identifies the need to balance three distinctly different needs suggests that, 'HRM is comprised of a variable mixture of three forms of management practice' (Torrington and Hall, 1987), which includes (i) conventional operational personnel management; (ii) the generic responsibility of line managers for day-to-day people management; and (iii) the strategic business policy decision-making activity designed to ensure a coherent and integrated approach to the overall management of the organization.

6.3.2 The change in emphasis

One of the more fundamental questions to be answered is why this change of emphasis from 'personnel administration' to 'HRM' was perceived to be necessary. Was this simply a new fad, or did this change demonstrate a significant shift in the way in which people were being viewed within organizations and therefore a reflection of changes which were occurring in the wider business environment? The answer is related predominantly to the latter trend.

It is interesting to note that the shift towards HRM straddled both a major boom period in the UK with its emphasis on 'strategic resourcing' and an equally significant, and deep, recession with its focus on 'downsizing' and 'rightsizing'. Economic pressures in both cases have emphasized the importance of, and costs of, people. In boom times this is often associated with difficulties in recruiting staff and the need to reward and recognize achievement internally within the organization. In periods of recession the pressure has been more related to the need to improve productivity (e.g. through the use of Total Quality Management, TQM), or enhance flexibility and skills development.

A secondary question which also requires some explanation is why 'traditional personnel management' was not perceived to be fulfilling the need for a more strategic approach to managing people. The answer to this question lies partially in the historical development of personnel management and the related change in the power and political influence of personnel in organizations. Torrington and Hall (1987) suggest that there have been at least five stages in the development of personnel management.

1. Welfare officer: the early development of personnel was associated with the formation of the Institute of Welfare Officers in 1913. This original role, as devised by a number of paternalistic employers, was to dispense benefits to 'deserving and unfortunate employees'.
2. Humane bureaucrat: as organizations increased in size and specialization and management science developed, the need developed for specialist 'bureaucratic' support functions in organizations. Issues such as job descriptions, job evaluation, the management of employee benefits, training and so on became more important. The 'bureaucratic' role therefore became a significant feature of the personnel function.
3. Consensus negotiator: after the Second World War, skilled labour became a particularly scarce resource and trade unions extended their membership. This led to the third role. In this newly emergent role the personnel function was charged with negotiating with unions representing employees.
4. Manpower analyst: as business developed through the 1960s and the need for 'manpower' became more acute, a fourth role developed – that of 'manpower analyst'. Manpower planning, to a large extent based on the assumption of organizational expansion, became a more significant responsibility. Generally, such planning was also based on the premise that the future would look very much like the past, albeit with greater requirements on staffing. The rapid rate of change which we have seen in recent years has been a significant challenge for this functional element of personnel/HRM.
5. From personnel to human resource manager?: while all of the roles and responsibilities outlined above are also representative of HRM, these roles and activities are primarily designed for an environment which is evolving in a gradual manner rather than one which is experiencing dramatic and unpredictable change.

One of the main criticisms of the personnel function in the past decade has been its inability to respond to the rapid rate of change which has affected business and its perceived lack of responsiveness to these changes. Many of the traditional roles and responsibilities, while still significant, are becoming of lesser importance in comparison with other requirements such as:

- facilitating and managing change;
- managing internal communication;
- developing the skills and competencies of all staff;

- devolving 'people' responsibility to line managers; and
- linking these activities to the strategic business objectives of the organization.

The critical skill is, however, to ensure that the administrative activities are managed in an efficient manner while at the same time developing these other needs listed above. In this latter guise the role of those charged with such activities is to act as agents and facilitators of whatever 'change' the partnership/board wishes to implement.

6.4 HRM AS AN AGENT OF CHANGE

Sections 6.2 and 6.3 have emphasized the changing nature of HRM and the specific nature of such activities in PSOs. To provide a basis for considering this changing role in more detail, a model of HRM activities is illustrated in Figure 6.1. This is considered in more detail below.

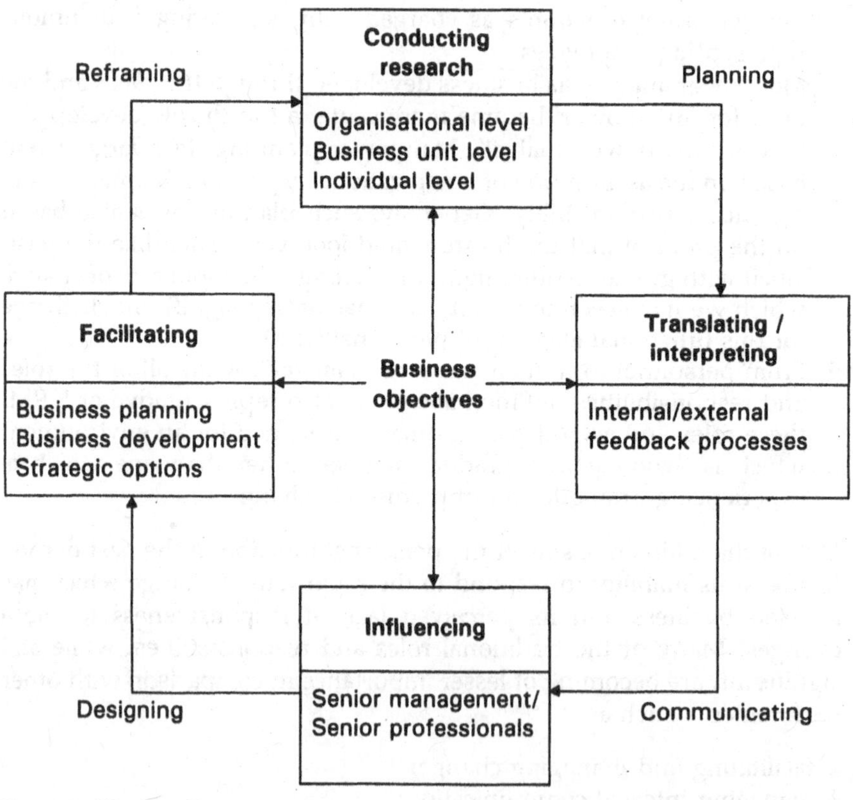

Fig. 6.1 Human Resource Management as a potential agent of change.

At the heart of the HRM model are the business objectives of the organization. Without a clear understanding of these, and of the 'people' consequences associated with such objectives, it is difficult – if not impossible – for those operating in this role to be fully effective. In some instances, however, the business objectives may either be expressed in very general terms or may not be explicitly defined. In such cases the role of those operating in the HRM field may be to help clarify the objectives and make them more explicit. In either case the skills required include:

1. An ability to **conduct research**. Research in this context refers to data gathering and analysis to help senior managers understand trends occurring at an organizational, business unit level or individual professional level. Some recent examples of this approach are: the use of questionnaires to gather information on variations in organizational culture before a recent merger; the use of Strategic Options, a questionnaire which helps a business unit to clarify the business strategies it is currently following and the options available for the future; and the use of focus groups and questionnaires to gather data on the views of senior personnel about a range of alternative reward strategies.
2. A capability to **translate and interpret** such data in order to inform senior management decision making. In this case the role of those involved with HR matters may be to help define some of the business dilemmas facing the organization or business unit. Some examples to illustrate the nature of these dilemmas in PSO might include: (i) the tension between local accountability and the need to retain central responsibility for setting the overall context within which the practice should operate together with control over major strategic decisions; (ii) related to the first dilemma, the need to encourage independence in business development terms while at the same time ensuring that inter-dependence between business units is fostered; and (iii) the importance of encouraging people to challenge the status quo without at the same time losing that which is effective in the current structure, in some cases referred to as the tension between 'corporate navel gazing' and 'corporate fee earning'!
3. An understanding of the importance of **influencing** senior professionals (as compared with either simply informing or demanding change). To operate in this role requires mutual trust and understanding, a broad appreciation of the nature of the issues facing the business together with effective communication skills. Above all, it requires the use of multiple networks both within and external to the organization in order to help make sense of, and validate, the patterns (if any) which emerge from the research exercise. Increasingly, it will also involve devolving responsibility for HR matters to

line managers. Education and training have a vital part to play in this process, particularly in the field of employee relations. Influencing professionals so that they become more confident and skilled when dealing with staff matters – be they of a motivational or disciplinary nature – is a significant feature of the new role which HR can play within the organization.

4. Linked to (3) is the ability to **facilitate** change by becoming actively involved in the change process. An example of this approach might include becoming involved in business planning/development workshops. This may also inform the research required in order to better define the nature of the changes required within the organization.

6.4.1 Conclusions

This paper has considered the nature of the changes affecting personnel/HRM activities within professional practices operating in the land, construction and property advisory services sector. None of the changes outlined in the paper is unique to this sector of business; indeed, these and other changes are having a considerable impact on all organizations. What is different, however, is the unique cultural setting within which these changes are occurring. Professional Service Organizations are different from what are euphemistically referred to as 'traditional hierarchical' organizations. For those charged with responsibility for the 'people' side of such businesses it is vital that these differences are fully appreciated.

In view of the changes new skills are required, skills which are quite different from those traditionally associated with that of 'staff', 'managing' or 'Senior' partner. Nor are they the traditional skills of the personnel professional. The challenge for the future will be to enable those who operate in this arena to develop the portfolio of skills demanded by this vitally important area of practice management.

6.5 INTRODUCTION TO STRUCTURED TRAINING

Concepts and terms such as continuing professional development, lifelong learning, vocational education and training have become commonplace in the professional and technical press since the introduction of mandatory rules and regulations by the professional institutions. The concept of lifelong education and training is not new, classic Greek writers such Plato and Solon identified learning as an on-going process.

Education and training of those working within the sphere of the built environment is essential for both employee and employer alike in order to meet the needs of an ever-changing industry. The RICS *Guide To Continuing Professional Development* states, 'Professional men and women have always accepted that learning lasts as long as their careers: it does not end with qualification!'.

This statement would appear to concur with much written on the subject of lifelong learning for committed professionals but rather than evaluating the educational theory of this subject, I wish now to examine what actually occurs in practice. This section will attempt to consider the practicalities and problems associated with education and training in the workplace using the initial findings of research into Continuing Professional Development (CPD) for Chartered Surveyors as case study material. It will also examine the relationship that exists between employees, employers and professional bodies with reference to education and training and the benefits that a structured relationship could bring to both individual and organization alike.

The research carried out by Simon Murray of the University of Northumbria into professional education and training is based on Chartered Surveyors in the Northumberland and Durham branch of the RICS and involved the distribution of a two-page questionnaire and subsequent analysis of the data provided. Of a total branch membership in excess of 1700 surveyors there are approximately 1300 members in the branch that must meet the mandatory CPD requirements of the institution and all of these were issued with a questionnaire. The total number of completed and useable submissions was 248, representing a response rate of 19%

6.6 THE NEED FOR EDUCATION AND TRAINING

One of the first questions that individuals or organizations will ask is, 'Why is training needed?'. Before considering the need for professional training it may be of use to consider possible problems associated with companies that spend little time or attention on training and educating their workforce. Whetherly (1994) states that such organizations may exhibit the following symptoms:

- Failing to meet organizational objectives.
- Change is difficult to achieve because staff have got out of or have never developed the habit of learning at work.
- A high rate of staff turnover, especially of the able and ambitious.
- Underdeveloped staff who are not ready for promotion, feel stuck in the organization or are capable of competing for jobs outside.
- Demotivated and disillusioned staff.

If any organization exhibited one or more of the above symptoms they would not be as efficient or cost-effective as possible and this may inevitably lead to a reduction in the services provided by the company, poor staff morale, reduction in available resources, crisis management decisions, reduction in demand from clients and – if not addressed and altered – ultimate failure.

Training can help in the development and growth of both employee and employer alike and a planned and structured staff development and training policy is essential. Whetherly goes on to state that such a policy is likely to produce the following results in an organization:

- Objectives are more likely to be achieved if staff are trained.
- Staff motivation is high.
- Development of services provided by the company is more likely if staff have the necessary attitude, knowledge and skills needed for change.
- A quality service can and will be supplied to clients/customers.
- The organization is ready for the changing requirements of the market place.
- Quality services are provided.

Therefore it can be argued that employers gain greatly from an educated and trained workforce. Payne states that what most employers want is a change in attitude and practice in the work environment and a greater degree of flexibility of their employees. If this is to occur then employers must have a proactive policy with regard to the training of staff, this must include both the financing of and 'freeing up' of staff time for training. Peters adds to this by saying that companies must consider training to be the research and development expenditure for the company and that it should be funded with at last 4% of gross revenues.

As far as the specific needs of the individual are concerned these will be dictated by the company that employs them, the professional institution to which they belong, and the personal motivation that they have towards their professional development. The growth of confidence and self-esteem of the individual is an important factor in effective employee development schemes but on the negative side many employees also view training as a waste of time, which is a rather disappointing and short-sighted attitude.

Professional institutions also have an important input into the education and training of their membership as most, if not all, expect formal compliance with their rules and regulations with regard to professional development. As well as providing a regulatory framework, they also tend to be one of the largest organizers of formal professional education and post-qualification training events.

6.7 THE RESULTS

As previously stated the response rate to the request for information was 19% of the branch membership. This rather low response may be

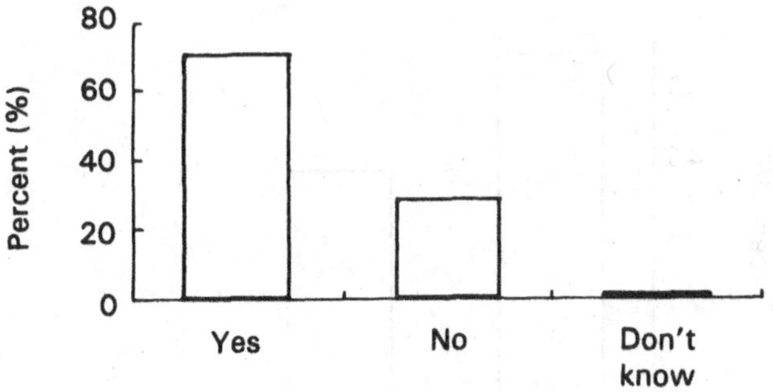

Fig. 6.2 Professionals' knowledge of the rules of CPD.

indicative of practitioners' views of continuing professional development, or may just be a function of the number of different questionnaires appearing on the desks of busy practitioners! The first question asked related to whether members were fully conversant with the Institution's rules with regard to CPD (Figure 6.2).

Approximately 70% stated that knew what they were and 30% stated either they did not know or were unsure of the rules governing CPD. One should also remember when evaluating these data that these are the members that took time to complete and return the questionnaire and could be regarded as those that take an active interest in professional education and training. The next question was a 'follow-up', asking those who replied 'yes' to Q.1 to state what they were. The purpose of this was to clarify whether those that stated that they knew what the rules were actually did. Of those that answered 'yes' to the first question, 36.29% did not know what the regulations were. When combined with the answers from Q.1, only 44.70% of the respondents appear to know exactly what is required by the RICS rules governing CPD for members. This would appear to suggest a lack of knowledge and understanding on behalf of members and that detailed explanation and guidance is needed.

The next data provided by the respondents were whether they did or did not comply with the regulations for CPD (Figure 6.3).

Of responders, 55.65% stated that they fully complied with the regulations, 38.71% did not and the other 5.64% do not know whether they did or did not. When compared with the 70.16% who stated that they knew what the requirements were (from Q.1) there is a 14.5% reduction in those that state that they meet the institution's rules and regulations. Nearly half of the respondents admitted that they do not meet the CPD requirements of the institution and this information was provided by the 19% who took time to complete and return the questionnaire.

The next information provided was general details about the type of

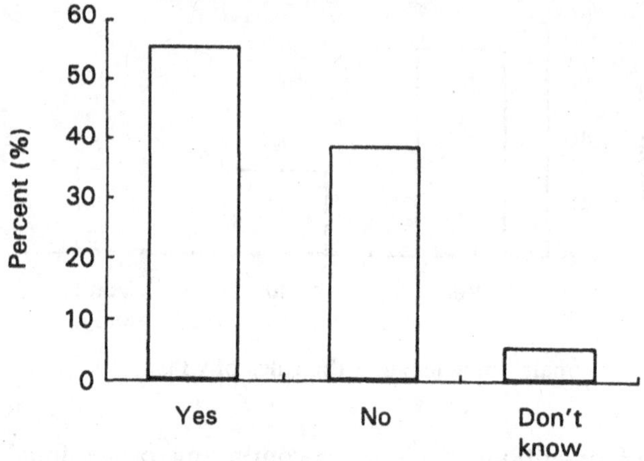

Fig. 6.3 Compliance with CPD regulations.

training that was carried out. The majority stated that their professional education was a combination of courses, seminars and private study, but when examined with the personal opinions and views given later in the questionnaire it would appear that there is a degree of confusion over what is and what is not CPD. Views provided also suggest that a great number of chartered surveyors see professional education and training as a system of points or hours accumulation rather that a method of learning and personal professional development. An interesting point of information to come from this section was the amount in-house training provided; only 17% of respondents stated that they attended events organized by their employers.

The next series of questions were designed to identify members' views of the existing management and regulatory procedures for CPD. The first of these questions asked if a formal policing system was needed (Figure 6.4).

Approximately 60% of respondents felt that a formal system was not needed. The next question then asked if such a system was introduced did they believe that members attitudes and compliance with CPD would alter (Figure 6.5).

As can be seen in Figure 6.5, over 60% felt that there would be a change in the attitude of members. When considered with the specific comments provided by respondents it would appear that the majority of members want a self-regulatory system but state that if a more formal system were in operation, their activity and attitudes would alter. The major issues raised by this aspect of professional training which need to be addressed by all concerned parties are the questions of whether self-regulation works, what would be the size and cost of operating a well-managed formal system, and would a formal system help produce a

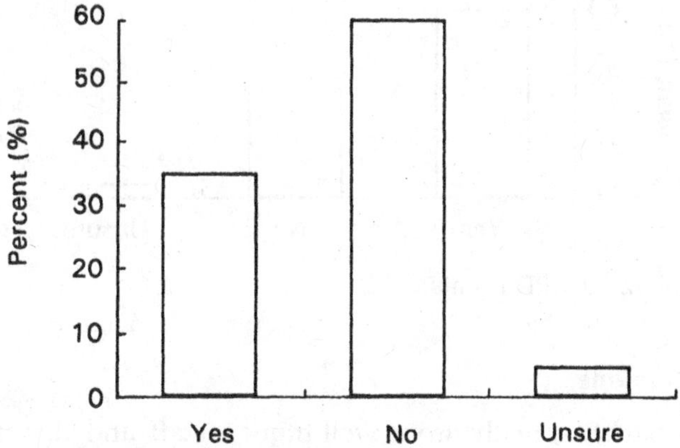

Fig. 6.4 Policing of CPD regulations.

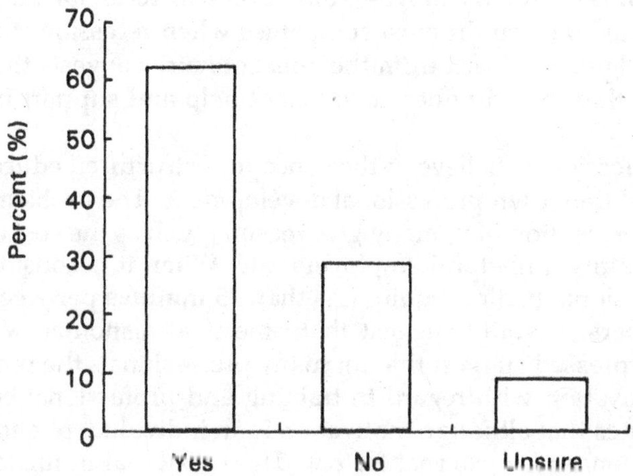

Fig. 6.5 Attitude towards CPD regulations.

more educated, informed and trained profession than already exists anyway?

The next two questions sought the views on how organized CPD and training events should be financed and paid for. The first of these questions (Figure 6.6) asked if members should pay for such events. Almost 60% of respondents felt that they should pay for organized events that they attended and approximately 38% felt that they should not. Those that answered 'no' to the above question were then asked to suggest ways on how such events should be funded. The two most common answers were from annual professional subscriptions and by active financing and promotion of events by employers.

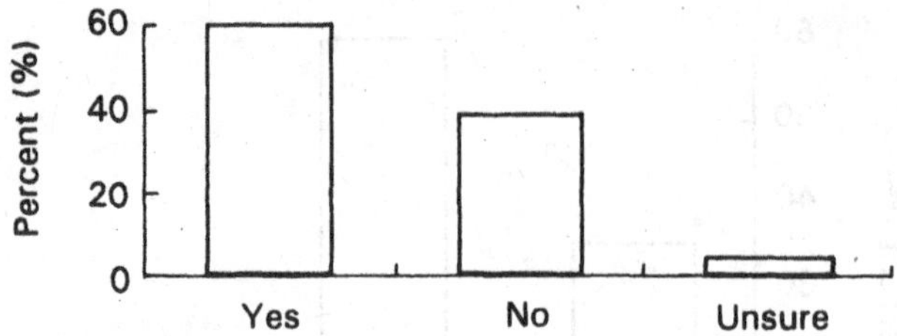

Fig. 6.6 Paying for CPD events.

6.7.1 Conclusions

Employers benefit greatly from well-trained staff and therefore they should also contribute to the training process. Some employers in the construction industry and its related professions have excellent staff development and training systems, but most do not. Construction is an industry which regularly moves from booms to recessions and one of the first items to be cut in most companies when recession starts is the training budget. Feedback from the questionnaire suggests that professionals feel that they do not get sufficient help and support from their employers.

Most professionals believe in the concept of structured education and training and their own professional development. The problem appears to be in the execution of it; many give reasons such as the cost of events, time constraints, unhelpful employers, etc. When it is considered that most professional bodies require less than 25 minutes per week CPD of their members, I would suggest that time is a misnomer. Views and opinions expressed suggest that for many professionals the problem is a lack of motivation with regard to training and professional education. Mitchell states that although motivation is an individual phenomenon it can be influenced by external factors. These external influences to the motivation to train must be positive and should be provided by employers and professional bodies.

Professional institutions have created regulatory frameworks and systems to control, influence and monitor professional development of their members. The philosophy behind such systems is correct in that one of the prime objectives is to promote professional development and excellence through education and training. The major problem appears to be that a high percentage of professionals are failing to meet their CPD obligations in the belief that no action has or will be taken.

The term 'lifelong learning' includes formal and professional education, workplace training, personal career development, etc. and all are part of this concept. To ensure that lifelong learning is successful you must have effective lifelong learners. Universities and colleges have now structured courses that facilitate a deep approach to learning rather than a surface or rote approach. Most employers nowadays want 'problem-solvers' and 'thinkers' with a understanding of the basic skills rather than just 'technical doers'.

Education and training of professionals throughout their careers benefits all concerned. It links individuals, employers and professional institutions engaged in the development industry and is essential in order to meet ever-changing needs of the industry and clients alike. Therefore a balanced and equitable learning partnership between all parties is crucial and fundamental to ensure success.

6.8 REFERENCES AND FURTHER READING

Annett, J. and Warr, P.B. (eds) (1981) *Psychology at Work*, Penguin, Middlesex.

Candy, P.C. (1990) How People Learn to Learn; Learning Across the Lifespan (eds R.M. Smith *et al.*), Jossey-Bass, Oxford, pp. 30–63.

Coulson-Thomas, C.J. (1992) Integrating Learning and Working. *Education and Training*, **34**(1), 25–9.

Cropley, A.J. and Dave, R.H. (1978) *Lifelong Education and the Training of Teachers*, Pergamon Press, Oxford.

Crowder, M. and Pupyin, K. (1993) *The Motivation to Train*, Department of Employment, Research Series No. 20.

Downie, R.S. (1990) Professions and Professionalism. *Journal of Philosophy of Education*, 24(2).

Guest, D. (1989) Personnel and HRM: can you tell the difference? *Personnel Management*, **147–59** (January), 48–50.

Handy, C. (1985) *Understanding Organizations*, Penguin, Middlesex.

Henry, C. and Pettigrew, A. (1986) The practice of strategic Human Resource Management. *Personnel Review*, **15**(5), 3–8.

Jarvis, D. (1983) *Professional Education*, Croom Helm, London.

Maguire, M., Maguire, S. and Felstead, A. (1993) *Factors influencing individual commitment to Lifelong Learning*, Department of Employment, Research Series No. 9.

Middlehurst, R.M. and Kennie, T.J.M. (1995) Leadership and Professionals: Comparative Frameworks. *Tertiary Education and Management*, Vol. **2**, August.

Mitchell, T.R. (1982) Motivation: new directions for theory, research and practice. *Academy of Management Review*, **7**(1), 80–8.

Payne, J. (1993) Too little of a good thing? Adult education and the workplace. *Adults Learning*, **10**(4).

Peters, T. (1994) *The Pursuit of Wow*, Macmillan, London.

RICS (1994) *Education Policy – A Strategy for Action*. Discussion document produced by Education and Membership Committee.

RICS (1988) *Guide to Continuing Professional Development*. Regulations made by General Council, 1 February 1988.

Torrington, D. and Hall, L. (1987) *Personnel Management: A New Approach*, Prentice-Hall, London.

Watkins, J., Drury, L. and Preddy, D. (1992) *From Evolution to Revolution: the Pressures on Professional Life in the 1990s*, University of Bristol.

Whetherly, J (1994) *Management of Training and Staff Development*, Library Association Publishing, London.

CHAPTER 7

Investing in people

7.1 PRACTICE MANAGEMENT SYSTEMS

7.1.1 Practice management

Most professional service organizations have evolved day-to-day office systems and administrative procedures to allow them to carry out the tasks and roles required to provide the service to their clients. But is just a set of standard procedure enough for them to provide a 'quality' service to these clients? Perhaps in the past the professional firm concentrated too much on the job in hand, and little business planning was carried out in service development. Much business development, if it happened, may have been spontaneous, intuitive, fortuitous, individual or reactive. Clear, firm-wide commitment to agreed objectives perhaps occurred only when the firm was threatened by external pressures or market changes.

No-one operates in isolation of the wider economy. Any organization must constantly consider the external influences on their business. Change is now a fact of life as markets are contracting or developing at an ever-increasing rate. Clients and staff are more demanding. Tendering and competition in the professions is now the order of the day rather than being commissioned as in previous years.

7.1.2 Styles of practice management

Each Practice is unique. There are perhaps two extremes in the styles of practice management:

1. **Traditional**: the traditional Practice sets future targets based on historical performance and, therefore, historical client decisions. It imposes a 'command and control' style of management based on conventions. Change is usually reactive and not well thought through. Training is poorly planned. Technically skilled professionals are rewarded with management positions for which they are untrained. The firm is administered rather than focused.
2. **Strategic**: the strategic firm sets targets based on what clients wish to achieve and a vision of the future. Staff know the Practice goals,

understand the changes to come and the skills that will be needed. Sound, inherent performance of people rather than rules are developed. Sustained profitability is achieved through strategic thinking and management skill.

Each Practice has its own view of itself and its own culture and management style.

7.1.3 Formal management systems

Each Practice should organise and control all matters that it perceives would affect its performance. Few 'evolved' management systems are fully effective and coordinated, and research has shown that most professional firms have only 20–40% of the essential management skills and disciplines required of a modern business organization. This chapter will highlight a basis for assessing individual Practices in terms of quality systems, given that work is carried out by expert staff who trade on the basis of their knowledge and skill. Many firms are changing, as they increasingly think through and design relevant Quality Management Systems. A formal system should simply document a 'mutual code of conduct' that confidently sets down in sharp focus what is to be done and by whom in order to achieve the aims of the firm and their clients.

7.1.4 Styles of Quality Management Systems

There are essentially two extreme kinds of formal Quality Management System:

1. **Traditional**: the traditional Quality Management System is typically procedural and sets no future targets. It often imposes technical conventions and a command and control style of management.
2. **Strategic**: this form of Quality Management System sets targets based on a vision of the future and what the firm wishes to achieve. Sound, inherent performance of people is developed rather than rule books.

In addition to day-to-day procedures, many firms are now formally addressing clear business aims and the commercially relevant professional development of their staff. Figure 7.1 shows a simplified diagram of the modern shift in professional practices from a traditional procedure-based organization to a much more strategic 'people-based' organization. This shift has major implications for the way the firm operates and is managed, since it is the responsibility of the Partners/ Directors to guide the firm in the chosen strategic direction. For example, if the firm wished to stay in the Now/Procedures box, they are effectively saying that they are a traditional firm concentrating on

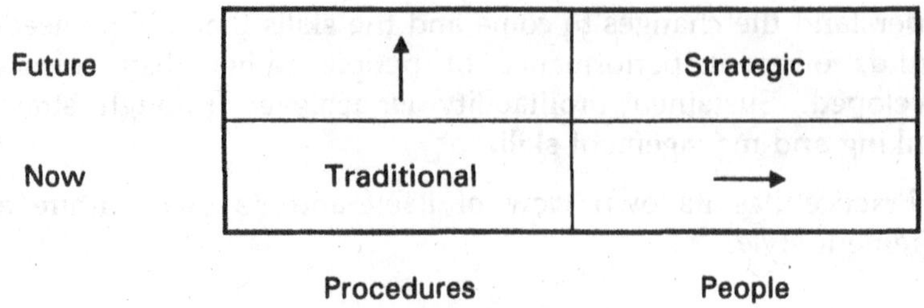

Fig. 7.1 Strategic focus of business planning.

traditional techniques, and would market themselves accordingly. Moving into the Now/People box would allow them to take advantage of the wider skills of the professional staff of the firm, thus requiring a slightly different managerial philosophy (see Chapter 9). A conscious decision to move into the Future/Procedures box requires an analysis of the way that the marketplace is changing and developing updated procedure to suit. However, the firms which are really addressing the future of both the industry and society are looking at the future marketplace and servicing it with people skills rather than procedures, which will enable the flexibility of operations required in a rapidly changing world.

7.2 'INVESTORS IN PEOPLE': THE PRINCIPLES

7.2.1 National management standards

National management standards such as the Quality Assurance standard BS/EN/ISO 9000 and 'Investors in People' provide a bench mark for measuring the effectiveness of a business as assessed against a series of predetermined indicators. A formal Quality Management System that complies with the National Management Standards will have considered and addressed in some way the commonly accepted good management practices including agreed business objectives, organizational structure, professional practice and routine administration.

Both styles of management mentioned earlier comply with the current National Management Standard for Quality Assurance although this is likely to change by the year 2000.

7.2.2 The aims and benefits of being an 'Investor in People'

The Investors in People standard also recognizes the more strategic organizations who motivate and develop their staff as an essential part of achieving their own clear business aims. The aims of an Investor in People are twofold:

1. To maximize the potential of the business.
2. To maximize the potential of individuals.

The benefits of being an 'Investor in People' can be:

1. clear business focus and awareness of market needs;
2. clear business goals and objectives;
3. effective organization of roles and responsibilities;
4. effective business and departmental communications;
5. competent and motivated people at all levels;
6. effective individual performance;
7. improved competitiveness; and
8. improved profitability.

7.2.3 The Investors In People standard

The principles of the Investors in People standard are in four parts:

1. **Commitment**: a commitment to identify and communicate to all personnel the aims and objectives of the firm and their contribution to achieving these.
2. **Planning**: training and development is planned to achieve business aims and objectives.
3. **Action**: training and development is given to all personnel in order to meet the business aims and objectives.
4. **Evaluation**: the investment in training and development is reviewed for success in achieving business goals and targets and to achieve further improvement.

7.2.4 The Investors in People indicators

The four standards are supported by 24 indicators (two are optional). The indicators are indistinguishable from sound business planning, training, motivation and leadership, and are used by an Assessor to ascertain compliance with the standard. The indicators are set down in the National Standard available from Training and Enterprise Councils (TECs) or in Scotland, Local Enterprise Councils (LECs).

The consultancy 'Investors in Quality' have interpreted the indicators into a questionnaire/diagnostic survey that will help the managers of a professional service firm to ascertain their current compliance with the IiP standard. Table 7.1 gives a diagnostic questionnaire developed by Chris Hobbs and Investors in Quality, which, when completed by staff at all levels within a professional practice, can be summarized to give the firm a good indication as to the level of current compliance.

Table 7.1 Investors in People; diagnostic questionnaire

How do we compare?	*Partners/ directors*		*Fee earners*		*Support staff*	
Statements:	*Yes*	*No*	*Yes*	*No*	*Yes*	*No*
I can clearly explain what the business or my team wishes to achieve in the future and how I contribute.						
We all know what we need to do to be effective in our jobs.						
Communications are good. I am always kept informed about matters affecting me and others understand my job.						
When change occurs in the structure, personnel, practice or procedures these are explained effectively.						
All new people, or people who change jobs are briefed effectively, trained and competent to do their new job.						
The business is committed to involving, training and developing people at all levels so as to achieve its aims.						
I receive effective feedback and review of my performance and know that concerns I have will be constructively addressed.						
We involve our clients in determining the standards of our service and changes are made to address their needs.						
This practice gets the best out of me and I know that my suggestions and proposals will be properly considered.						
Senior people in the firm are good leaders and managers as well as good professionals.						
Date: Totals:						

Compliance by the traditional professional firm

The diagnostic questionnaire is not comprehensive, but should give a good overview regarding current compliance with the National Management Standards. Generally, professional firms can demonstrate compliance with only about 10% of the Investors in People standards. The most common omissions or shortcomings occur in clearly defining or implementing the following:

1. business ambitions
2. roles and responsibilities
3. communications
4. recruitment and selection
5. induction of newcomers
6. feedback on performance
7. training and development

7.2.5 Investors in People accreditation

Companies complying with the standard can seek assessment by their local TEC/LEC. In preparing for assessment companies usually compile a portfolio or a directory indicating to the Assessor the documentation and evidence that is available to help demonstrate compliance with the standard. Documentation is not enough on its own. Interviews demonstrate to the Assessor whether the standard is ingrained as a habit, rather than fine intent or tick lists. On successful assessment the firm is accredited, and be allowed to use the laurel logo of 'Investors in People'.

7.3 THE KEY STAGES TO RECOGNITION

The key stages to external recognition as an Investor in People are shown in Table 7.2. They are set into five major stages (the five Ps). Accreditation for any nationally recognized standard of achievement is a complicated and expensive process, so the first stage of policy formulation is very important. It may be that the commitment to the process should be the first decision to be made, which may come from external driving forces such as client requirements or a competitor analysis. Following that is a situation analysis of where the firm is at the moment and where it wants to get to in terms of management standards. The Action Plan of how the firm is to achieve the standards is then established with any consultants who are working with the firm, and then we move into the second 'P', that of Planning and Procedure.

This could be the longest and most tortuous of the five stages, since it involves the adaptation of the firm's own management systems, which may also involve a complete change of culture among the staff who have to carry them out. Remember that with Investing in People, we are not merely writing down existing procedures, which we could have got away with for QA accreditation, but possibly adding to or changing systems and processes so that we are a people-led, rather than

Table 7.2 Key stages to recognition as an Investor in People

Policies:	State your standards	What is the aim?
	Diagnostics and interviews (questionnaire)	How do we measure up to the standards?
	Action plan	What must we do?
	Commitment/policy	Are we committed?
Planning and procedures:	Development of practice management systems	Do we have the systems and tools?
Practice:	Communication	Are people informed?
	Implementation/audit	Are things happening?
Performance:	Review/evaluation	Were the aims achieved?
	Develop evidence	Can we prove it?
Proof:	Pre-assessment audit	Are we ready?
	Assessment	Are we there?
	Accreditation	Bingo!

procedure-led, organization. The development of any new systems must therefore be very carefully thought through.

Once the firm has decided what systems and procedures suit our organizational structure and culture, we must both communicate these to all concerned and – most importantly – implement them together with any criteria for assessing their operation and effectiveness. After some time in operation we must review and evaluate whether the aims have been achieved and, if all is going well, progress to formal accreditation.

7.3.1 Summary

Owing to the external pressures of the recession, increased competition, and reducing professional fees, many firms are now questioning the way that things are being done and changing from traditional ways of managing their firms to more flexible ways, which will give a greater opportunity to service new and developing markets. They increasingly realize that their objectives and plans should be based on the future needs of their clients rather than on historical conventions.

All businesses need positive plans in order to achieve clear business goals, whether they are in the business of manufacturing, traditional services or professional services. In later chapters of this book, we will discuss whether professional services are, or should be, a business, but at present let us make that assumption, and state that all businesses need to keep an eye on the changing marketplace and adapt their own systems accordingly. Investors in People is a process which allows firms to progress through the Quality Assurance phase, to Total Quality Management to the new concept of the 'Learning Organization', and encourages firms to value the people within as the highest priority.

Leaders need to establish and state, in conceptual terms – as well as keep under constant review – what their clients want and what their staff will need to be able to do in the future.

The Investors in People standard can provide a firm with a target or bench-mark against which to assess and measure itself as compared with the good management practice of many other firms. The benefits all make good business sense.

7.4 PRACTICAL ILLUSTRATION

7.4.1 Case study: Drivers Jonas

- Size of firm: 270, including 37 partners.
- Accreditation as an Investor in People achieved in the latter half of 1994.

Brief outline of the practices and services

Drivers Jonas is an independent private partnership of chartered surveyors and international real-estate consultants, founded in London in 1725.

Offices are in: London; Nottingham; Norwich; Manchester; Glasgow; Berlin; Toronto

Professional services include: Agency; Management; Investment; Valuation; Property management; Building surveying; Commercial development; Land-use planning; Economic development; Urban regeneration; Environmental assessment.

Reasons for undertaking the IiP initiative

IiP became part of a major programme of management change, started in 1992, designed to improve performance and quality at all levels and in all aspects of the business.

Initial diagnostics and key findings

The diagnostic survey was carried out in late 1993, by which time most of the initiatives arising from the management change programme were in place. As a result, the findings from the diagnosis indicated only a few actions needed to be taken to meet the IiP standard. These were to:

- improve induction of new staff at departmental level (it was effective at corporate level);
- improve content of in-house magazine to make written communication more up-to-date, immediate and relevant;
- ensure secretarial and administrative staff knew the business goals and performance of their teams; and
- review and improve evaluation processes.

Costs

- **External**: 4 days consultant's time for diagnostic phase and action plan; 6 days assessor's time during assessment.
- **Internal**: 40 man days over a 6-month period during implementation of the action plan.

Actual benefits

- Improved employee focus and awareness of their role and contribution.
- Higher levels of morale and motivation.
- Increased individual and team performance.
- High quality of staff and service to clients.
- Enhancement of reputation with clients. Several clients had themselves achieved accreditation or were working towards it and were quick to congratulate, other were keen to be told what IiP meant.
- Improved competitiveness.

CHAPTER 8

Managing teamwork and leading professional people

8.1 INTRODUCTION

It has long been acknowledged that construction is a team process, since it involves the specialized input of practitioners from a number of diverse disciplines to enable it to succeed. Indeed, as the construction process has become increasingly more complex, then the need for a multi-disciplinary team to undertake the development of projects from inception to completion has become more acute.

Traditionally, the project 'team' has been segregated into designers, constructors and suppliers who are bound together by formal contractual ties which, at times, have become so restrictive that relationships between them have been more adversarial than cooperative. Even within the 'design team' there has tended to be conflicts of interest between team members who have generally come from different disciplines, with each pursuing their own objectives sometimes at the expense of the overall project.

If we accept Fiedler's definition of a team, which is: 'a set of individuals in face to face interaction who perceive each other as interrelated or as reciprocally affecting each other and who pursue a shared goal', then it is clear that the management of teams participating in the process of construction requires quite a radical revision.

It is our intention in this chapter to concentrate on the management of the design team in particular, but the issues considered may also be applied to the management of the construction team.

8.2 TRADITIONAL APPROACHES TO MANAGEMENT OF THE DESIGN TEAM

The appointment of the design team has traditionally been the prerogative of the client. At the inception stage of a project the client generally engages the services of an architect, who then helps the client to develop a brief and, as the initial design begins to develop, advises the client in the selection of appropriate professionals from other disciplines to provide their specialist input. Traditionally, the disciplines from which these specialists have been drawn have been those of structural engineering, services engineering and quantity surveying, but with the increasing complexity of modern construction projects and the advent of meta-professions, mentioned earlier, other specialists have begun to emerge and may also be incorporated into the design team. These include the landscape architect, the interior designer, the fire safety expert, the energy conservation consultant and the acoustics consultant to name but a few.

It is also being recognized that contractors, or more correctly, construction firms – in particular those with a specialist knowledge of their specific area of construction expertise – may also have an important role to play in the design process and should therefore be incorporated into the design team.

Having entrusted the architect to assemble a multi-disciplinary design team, the client – who may not have in-depth knowledge of the design or construction processes – has traditionally asked the architect to lead the design team and frequently, once the design has been completed, to lead the construction team as the client's representative. In fact, the Joint Contracts Tribunal (JCT) forms of contract are written with the architect undertaking the 'project management' role, and this has become known as the traditional procurement route. This strategy has operated with varying degrees of effectiveness for some considerable time now. The issue for us now is whether the effectiveness of this process could be improved following a retrospective consideration of the functioning of the design team and the way it is led.

There can be no argument that the design team needs a leader. All teams need effective leadership if they are to be successful. What needs to be debated, however, are what qualities that leader needs to possess and what tactics need to be employed by that leader in order to obtain the best output from the team. This debate should not centre around whether or not the architect is the best person to lead the design team, but should be an objective appraisal of what the effective leader needs to provide in order that the objectives of the team can be successfully realized.

Unfortunately, it is often common practice within design teams for construction projects to appoint a leader based on technical ability alone rather than their ability in managing people. An excellent designer may not possess the required interpersonal, informational and decisional skills which are an essential requirement for a good team leader.

The design team on a typical construction project differs substantially from the design teams in other manufacturing industries, since the construction industry is organized on a project-by-project basis and so is both temporary and dynamic in nature. In addition, many of the team members are not under the direct control of the team leader or even assigned to the project on a full-time basis. This can lead to problems in establishing good working relationships between professionals who may not have worked together before.

Where team members belong to different organizations and are therefore not directly subordinate to the team leader, it can be difficult to motivate them to achieve the desired results, since the commonly perceived direct methods of motivation are not available to the leader.

A further problem occurs where team members are drawn from a number of different disciplines and conflict emerges between the individual objectives of each team member and the overall objective of the team itself. Furthermore, because the design of construction projects involves the overlap of inputs from the different disciplines, conflict can also emerge between the team members themselves. Personality clashes between different members of the team can also add to intra-team conflict.

Divisions between the professions to which members of the design team belong, have been maintained to a large extent by the institutions that have promoted codes of conduct, fee scales and standard documents defining the terms of appointment and scope of services offered.

In order to overcome these barriers to effective team working it is useful to review essential characteristics of effective teams and consider how these may be incorporated into the management of design teams for construction projects.

8.3 EFFECTIVE TEAM CHARACTERISTICS

Groups produce better results than independent individuals. They can produce 'groupthink' which will generate more alternative courses of action than is likely with individuals acting independently. The group also tends to eliminate inferior contributions, average out errors and supports creative thinking. This synergy enables groups in most situations to outperform their most accomplished member. Conversely, they do also tend to act more radically, and sometimes more rashly than individuals because of the ability to hide within a group decision.

However, for groups to function effectively they need to act synonymously, and one way to achieve this is by developing cohesiveness within the group. Cohesiveness is a function of the attractiveness of the group which in turn is related to its composition. The status, values, attitudes, abilities and interests of the group members imbue the group with its perceived attractiveness.

Another important element of effective groups is the quality of their leadership. The leader needs to provide focus, direction and stability to the group. Groups with effective leaders display direction, dedication and assertiveness. Groups with ineffective leaders lack direction, are indecisive, suffer from internal conflict and fall apart easily.

Teams can be distinguished from other types of group by their need for interaction. Their members need to be interdependent and need to cooperate and coordinate their actions to accomplish the task or meet the objective. Leaders of teams need to possess two sets of functional requirements. These are related to the team's task-related needs and its socioemotional needs. Leadership attributes associated with task-related needs include coordinating, initiating contributions, evaluating, information-giving, information-seeking, opinion-giving, opinion-seeking and motivating individuals. Leadership attributes associated with the socioemotional needs of the team include, reconciling differences, arbitrating, encouraging participation and increasing interdependence among team members. One individual may possess all these attributes or some teams may possess one individual who satisfies the task-related needs while another satisfies the socioemotional needs. Thus, leaders may lead the team in one particular situation but revert to the function of a group member in another. It is therefore important when selecting and developing teams not to place too great an emphasis on the leader, as this can diminish the importance of the contribution made by other team members.

An important function of leadership is to facilitate efforts for planned change. Some changes occur whether or not people initiate them, but change may be planned or resisted by the efforts of concerned individuals. Organizations may be considered as mechanistic or organic according to their responsiveness to change. The mechanistic type is highly controlled, emphasizes obedience to superiors, and can function well only under stable conditions. The organic type is more open and adaptive to change. An organization which overstresses the maintenance of stability and the limitation of uncertainty runs the risk of losing the potential for innovation. Movement towards a more organic structure can be made through the recruitment and development of talented individuals, the fostering of an environment that encourages individuality of ideas, the ability of the organization to be self-critical and the flexibility of the organization to be adaptive to change.

As we have mentioned in several chapters previously, organizations are constantly changing to meet the needs of modern industrial and commercial requirements. In the past people were required to work in close proximity to each other to have access to the necessary information. Today, the telecommunications revolution has made decentralization much easier. The corporation of the past was like a vast army with divisional and battalion commanders working to the general's plan. Today's corporation can be likened to a network of guerrilla bands; independent, free-ranging, flexible, alert and highly trained rather than well-drilled. The emphasis is now on intelligence and enterprise rather than on conformity and obedience. The team leader must become less of a supervisor, giving instructions, determining procedures and controlling processes and more of a facilitator, encouraging contributions, stimulating participation and emphasizing innovation. There are no limits to the ability to contribute on the part of the properly selected, well-trained, appropriately supported and, above all, committed person. Thus, the modest-sized, task-oriented, semi-autonomous, mainly self-managing team should be the basic organization building block.

This need for autonomy among team members is the essential difference between the constrained team of the past and the flexible team of the future. It is accepted that for teams to function cooperatively then members must surrender some of their individual autonomy for the common good, but to surrender nearly all individual autonomy in favour of external control reduces flexibility and adaptability to change and stifles innovation. If moderate autonomy is to be nurtured in teams then the style of the leader must also change to allow more freedom, less initiating structure and more consideration for the needs and aspirations of the team members.

8.4 CONTEMPORARY APPROACHES TO CONSTRUCTION DESIGN TEAM MANAGEMENT

Although traditional forms of contractual arrangement still predominate in the construction industry, there is now clear evidence of a move towards a wider range of procurement methods, which provide the parties within the contract with a more adaptable choice of methods for engaging in construction projects. In some cases, such as design and build, the contractor manages the design and construction processes, whereas in other cases, such as project management, the manager of the design and construction processes is an independent professional, working for the client. Such managers have come from most of the professions represented in the design and construction teams.

Although it is not the purpose of this paper to advocate any specific procurement or contractual method, there is a great deal to be said for having a well-trained manager as leader of the design and construction processes. An independent professional such as a project manager can create an environment where the design process can flourish and break down the barriers between those having design ownership so that everyone in the team is encouraged to participate fully. The advantages of professionals being familiar with the roles of others in the industry is not disputed and although there are dangers in removing specialism, a better product is achieved if the designers have an appreciation of all constraints in the design rather than just those in their own area. Indeed, the Latham report, while seeking a clearer definition of the roles and duties of project managers, suggests that a 'lead designer' – who may not necessarily be an architect – may be a useful professional to employ on certain projects. In fact, many leading retail organizations have now divorced the role of detail design and supervision from that of concept design on construction projects which have been undertaken for them.

Managing a successful project demands the skills necessary to control time, cost, quality and performance. In addition, total integration of the consultants who form the multi-disciplinary design team has to be an underpinning philosophy to managing the design process. The leader must give specific attention to three essential groups of factors; the environmental factors, the membership factors and the dynamic factors.

- **Environmental factors** are concerned with the way the leader organizes the work such as the time limits that are imposed and the cost constraints involved. The project design environment must permit and promote team development and establish a climate of integrity and trust among team members.
- **Membership factors** are concerned with the actual people in the team. There must be common values and understanding together with a balanced distribution of knowledge and skills. Selection of team members should be carefully carried out so that they are complementary and the team must be sufficiently resourced to achieve its objectives.
- **Dynamic factors** are concerned with the organization of the team and its leadership. The team leader should possess sufficient knowledge and skill to enable the team to develop under his or her direction. This will require the definition and clarification of team objectives, team members' roles, administrative procedures and decision-making procedures.

This revolution in design team management will be enhanced by improved communications between the team members brought about by CAD technology which can broadcast design information electronically between them. This can help to identify design disparities which have long been a prominent source of conflict.

8.5 CONCLUSION

The need for a more enlightened approach to the management of the construction design process is required if the productivity target of a reduction in real costs of 30% by the year 2000 is to be achieved, as suggested in the Latham report. Much progress can be made towards integrating the design team by breaking down the barriers between the different professional disciplines. This distrust begins in the professional education since all the various parties, builders, engineers, surveyors and architects are still predominantly educated separately. Calls for the establishment of a double degree with a common core curriculum for the built environment have long been advocated by senior personnel within the construction industry, but continue to be resisted by many professional and educational institutions. Meanwhile the continued training of future project managers in the skills of successfully leading construction projects appears to bode well for the future of the industry.

In addition there must be a more flexible attitude adopted by the professions regarding the contribution that can be made to the design process by an increasingly varied pool of contributors. Designing and constructing a new building project calls on the skills and expertise of many individuals. Recognizing and managing their contribution and the interfaces that exist between them will be vital to the success of the project.

8.6 REFERENCES AND FURTHER READING

Adair, J. (1986) *Effective Team Building;* Gower, London.

Andrews, J. and Derbyshire, A. (1993) *Crossing Boundaries,* Construction Industry Council, London.

Bales, R.F. and Slater P.E. (1955) Role differentiation in small decision-making groups, in *Family, Socialization and Interaction Processes* (eds T. Parsons *et al.*) Free Press, New York.

Burns, T. and Stalker, G.M. (1961) *The Management of Innovation,* Tavistock, London.

Fiedler, F.E. (1967) *A Theory of Leadership Effectiveness,* McGraw-Hill, New York.

Hill, G.W. (1982) Group versus individual performance: are N+1 heads better than one? *Psychological Bulletin,* **91**, 45.

Jay, A. (1993) Are you a New Manager? *Observer Magazine,* 28th March.

Latham, M. (1994) *Constructing the Team,* HMSO, London.

Peters, T. (1988) *Thriving on Chaos,* Macmillan, London.

Smith, G and Morris, T. (1992) *Exploiting Shifting Boundaries; Uplifting Architecture.* RIBA Strategic Study of the Profession, Phase 1, Strategic Overview, RIBA, London.

CHAPTER 9

Leadership styles in professional firms

9.1 INTRODUCTION

The advent of new procurement methods for construction work and the wider role of the professional adviser has led to a fundamental reappraisal of the various functions carried out by construction professionals. Indeed, as has been mentioned previously, the more sophisticated clients are now taking an active role in the project management process itself. This has correspondingly led to a considerable change in the organization of both the projects and the traditional professional practice, with the architect or engineer no longer automatically considered to be the project leader. The leader of the project can now rise from any of the participants depending on the procurement method, form of agreement, etc.

Using a classification of procurement based on project responsibilities, this chapter looks at the accepted theories of leadership, how they were applied and by whom in the traditional project organization, which style is more appropriate under the alternative procurement systems, and what effect this has on practice management. It concludes with a consideration of the ramifications for the continuing education of construction professionals.

The traditional procurement system for construction in the UK has been developed over the past 150 years to a point where all the parties engaged in a construction project have a very definite awareness of their own functions and responsibilities, are reluctant to cross the demarcation lines, and require that students of their discipline concentrate on the narrow skills which have tended to make this process self-perpetuating. This traditional system can be classified as 'single stage lump sum selective tendering', whereby the design process is or should be fully complete before the builder commences work and indeed tenders for the work on the basis of a pricing document which is generally a Bill of Quantities produced by the consultant quantity surveyor by measurement of the finished design. However, over the past 20 years in the UK many more forms of procurement have been developed and used which reflect the relative importance the building owner puts on to:

- total duration of project;
- cost certainty at the start;
- type of construction; and
- single or multi-point responsibility.

Additionally, the services offered by professional practices have widened considerably and for a practice of quantity surveyors, for example, these can now include total project management, development value analysis, insurance valuations, company analysis and market analysis to complement the more traditional role of project cost planning and control. This has led to a major shift in the skills required in a progressive professional practice, from the mainly procedural skills inputting into the traditional procurement system to a wider portfolio of skills necessary to allow this wider range of services to be offered.

Obviously, this will have major implications for the management of a professional practice and it is the purpose of this chapter to investigate aspects of leadership within a practice (as opposed to within a project) to see how the style of leadership has changed and what characteristics should be sought in potential future leaders of professional practices.

9.2 THE TRADITIONAL PROFESSIONAL PRACTICE

Until relatively recently, the professional practices in the construction and property industries were organized as partnerships, due mainly to the regulations of the professional bodies such as the RIBA, RICS, ICE, etc. The partners were jointly and severally liable for the debts of the partnership and this liability was unlimited. This created a very strictly controlled career progression since the existing partners would very seldom entertain inviting anybody into partnership unless they were assured of their technical competence. However, a partnership has an added complexity since the partners are simultaneously shareholders and directors with an function of directing and managing the business in addition to direct fee-earning work.

The organization of the Practice was consequently predominantly vertical or a line management structure with the technical responsibility still at the peak of the organization.

This organization structure did have some flexibility depending on the size of the Practice and whether the work to be carried out in the design stage or construction stage. The main point for us is that the service offered was mainly procedural and similar for all projects undertaken, with the task of the leaders being to ensure that the job was carried out in the required time and to the required quality. All of the people in leadership positions, from team leaders through Associates to Partners had grown with this system and been promoted through the various levels.

Let us now look at the theories of leadership to attempt to classify this system as a 'control' to allow us to consider the requirements of the new types of practice.

9.3 LEADERSHIP THEORIES RELATED TO THE TRADITIONAL PRACTICE

There are three major theories relating to leadership:

1. Trait theory
2. Style theory
3. Contingency theory

The first two adequately explain the type of leadership and the criteria for selecting potential partners or managers in the traditional practice as described above. However, as the Practice diversifies, the different qualities required for leadership must be recognized.

9.3.1 Trait theory

This theory rests on the assumption that the individual is more important than the situation the leader has to cope with. Therefore, according to the theory, there are certain traits or characteristics common in all successful leaders. However, on testing this hypothesis it was found that very few of the traits of any particularly successful leader were common across the board, which implies that successful leadership depends on the situation rather than the person. Therefore, this suggests that good leaders can come from a wide variety of sources and the traits that lead to success may differ according to the situation.

The major assumption of trait theory – that leaders are born and not made – favours an emphasis on selection rather than training and also favours those about whom the selector can identify with regard to their own past experience.

Let us now look at the 'traits' found by the various researchers in successful leaders. List 1 are those traits found in virtually all leaders from a variety of organizations and industries; List 2 are other characteristics mentioned which are not common.

List 1

- Intelligence
- Initiative

- Self-assurance
- 'Helicopter'*

List 2
- Enthusiasm
- Sociability
- Integrity
- Imagination
- Decisiveness
- Determination
- Energy

*Note that the 'helicopter' trait is the ability to rise above the particulars of a situation and perceive it in its relation to the overall environment.

This trait theory has been criticised by management theorists because it cannot be applied to leaders in all situations and industries. However, we are only concerned with leaders in professional practices in the construction and property industries, and must therefore ask ourselves whether this is the best way to classify and select potential leaders in construction practices. In practice the vast majority of the managerial selection schemes in these professional organizations work on some assumed, and often unspecified, trait basis, i.e. what traits are most effective under our specific conditions. As we have seen in the Introduction, for many years the conditions under which practices operated were relatively unchanged so the traits or characteristics of effective leadership could be isolated and established in potential promotees; however, as the situation and function of professional practices changes, this method of establishing leadership potential loses some validity.

9.3.2 Style theory

Firstly, we should remind ourselves of the difference between leadership and management. According to Adair (1988), management is the business of achieving the task, with the resources available (human, financial and physical) and within the organizational structure by the process of planning, controlling, organizing and evaluating. Leadership, on the other hand, has the same basic aims but is more related to the situation, i.e. the actual task to be done, the actual resources employed, and actual organization.

The leadership style theory assumes that employees will work harder (and therefore more effectively) for managers who employ given styles of leadership. These styles can be classified on a continuum ranging from a totally directive ('telling') to a more participative style (Figure 9.1). The effectiveness of the style depends to a great extent on the

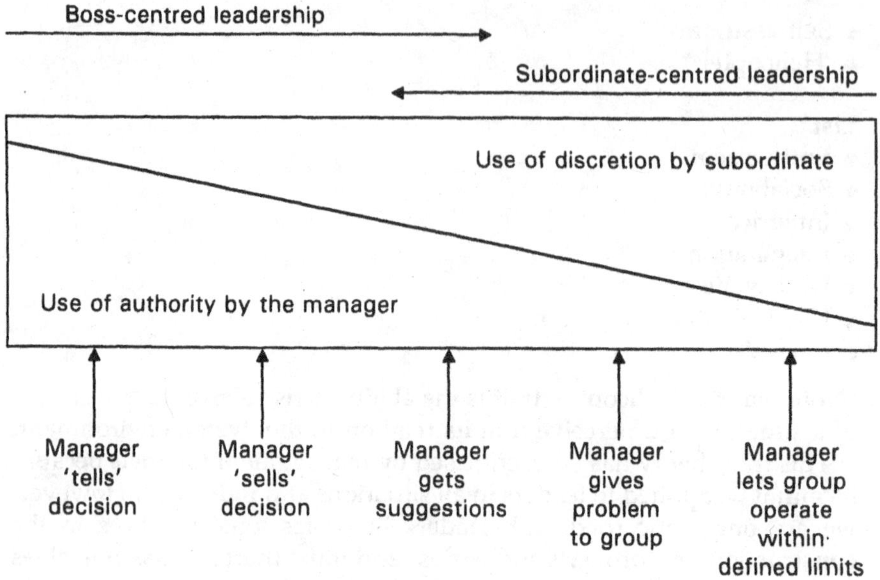

Fig. 9.1 Continuum of leadership styles.

subordinates and the amount of responsibility they are willing to take for the way that the task is done and obviously their personal relationship with the manager. As mentioned above, the traditional practice was very much a procedurally based organization so when a new project was started, the team put together was controlled by the process known as 'Management by Objectives' (MBO).

This implies that the traditional leadership style of a professional practice was one of a directive rather than participative style, with a higher concern for production than concern for people (Figure 9.2). This is backed up by the traditionally low salaries paid to younger and unqualified members of staff (even 'premiums' paid by articled clerks for the privilege of working for a Practice) and the historical poor quality of working accommodation.

There is evidence that supportive or participative styles of leadership, which are often preferred styles by subordinates, are related to staff satisfaction and lower turnover and grievance rates and indeed have been found to be associated with higher-producing work groups; however, we must be careful in such broad generalizations since it has been suggested that it could be more effective working that leads to (or permits) more supportive or participative styles. Also, some people prefer to be directed if they have low needs for independence (cf. Maslow's hierarchy) although professionally qualified people are, by definition, capable of and require, independence of action.

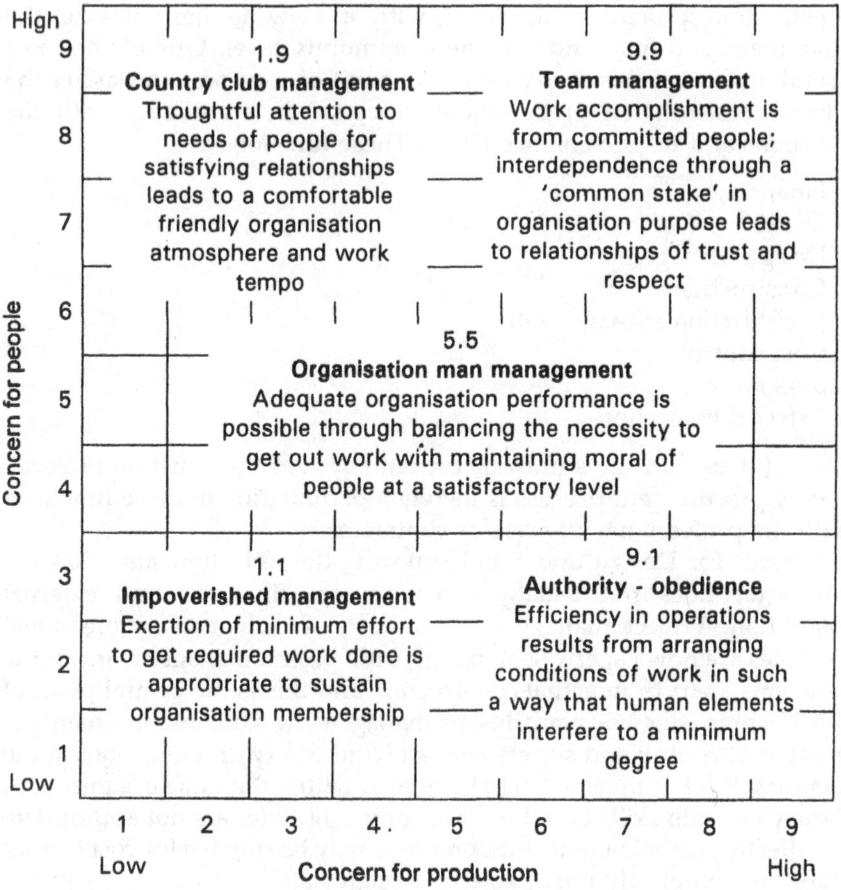

Fig. 9.2 The managerial grid.

9.4 THE EFFECT OF NEW PROCUREMENT METHODS ON THE PROFESSIONAL PRACTICE

The traditional 'single stage lump sum selective tendering' procedure now accounts for less than 60% of construction work in the UK when measured in terms of value. However, this may be slightly misleading since if measured in terms of the number of projects, it will be well above this figure. Nonetheless, clients who are becoming increasingly more sophisticated in terms of the construction industry are looking at the new arrangements with interest since the reduction in total project time which the new arrangements offer may translate into a substantial saving when rental values and interest payments are considered.

A study by Claude Mathurin in 1990 for the EC (now EU) looked at construction procurement, albeit with a view to harmonizing the guarantees and insurances at the Community level. One of the most useful aspects of this study from our point of view was to classify the various construction procurement processes in accordance with the functions and responsibilities taken. These functions are:

- Financing
- Design
- Design management
- Construction
- Construction management
- Cost control
- Supervision
- External inspection

Each of these functions must be carried out in all construction projects, so any procurement process is merely a permutation of these functions with the professional advisers or contractor.

Indeed, for Design and Build contracts the only functions that the contractor does not usually carry out are financing and external inspection. Project management contracts, which many professional practices are now eagerly seeking, give the project manager many of the functions apart from actual construction and this is the central point of this chapter, since design, design management, construction management, cost control and supervision all require very different skills at the technical level. If they are all to be offered within the same organization, then leadership skills based on traits or single styles are not appropriate for effective management since one style may be suitable for cost control staff, but completely unsuitable for design staff.

Additionally, the training required by staff with different skills and competencies will differ, so as continuing professional development becomes more important within the professions, the judgement of the leader regarding the training and updating needs of staff will weaken.

9.5 LEADERSHIP THEORIES RELATED TO PRACTICES WITH DIVERSE SKILLS AND SERVICES

9.5.1 Contingency theory

This contingency approach takes specific account of the other variables involved in a leadership situation, in particular the task and/or work group and the position of the leader within that work group. It is therefore much more appropriate when dealing with diverse technical skills within subordinates and takes more account of the motivational factors within professional work groups. Contingency theory does take into account the preferred style of the leader but also considers three other sets of influencing factors in any leadership situation, these are:

- the needs, attitudes and skills of the subordinates or colleagues;
- the requirements and goals of the job to be done; and
- the organization and its values and prejudices.

As mentioned above, leadership styles can be classified on a continuum from 'directive' to 'consultative', so we can now look at each of these four influencing factors to see which leadership style would best improve their effectiveness when considering a professional practice within the construction industry.

The leader

A particular leader's preferred place on the scale between directive or consultative will depend on seven major factors outlined below. Generally speaking, leaders in professional practices have attained that level by virtue of the 'Peter Principle' after having shone at a subordinate level; this is felt to be the major influence on the factors which tend to make the leader's style more directive. The factors are:

1. Value systems; how far does he or she feel that subordinates should be involved? Is he or she task- or people-orientated? (see Figure 9.2).
2. Confidence in subordinates; if the leader is more technically competent this will naturally tend to a directive style.
3. Habitual style; however much people try to adapt, it is difficult to teach an old dog new tricks.
4. Assessment of his or her own personal contribution; as a leader in their own field of expertise, a boss will usually tend to be directive, whereas in a field where they are less technically qualified, the style tends to be more consultative.
5. The need for certainty of outcome, i.e. how important is it to get the right answer, or at least a definite answer. The more definite the answer required, the less likely it will be.
6. The degree of stress to which the leader is subject.
7. The age of the leader.

The subordinates or members of the work group

Colleagues and subordinates in a professional office will have their own preferences for a tight or open response from the leader to any given situation, these preferences will be influenced by at least five factors:

1. Their expectations; do members of the group have ambitions for promotion and does a particular leadership style affect their perceived

chances of being 'spotted'. Is the group accustomed to having tasks assigned or do they organize and control their own work?

2. Their interest in the problem or situation; if the task is seen as routine and mechanistic then construction professionals tend to prefer structured or directive leadership.
3. Tolerance of ambiguity; information is very seldom passed vertically in a professional office, especially when a directive style is adopted. Therefore, if there is risk and uncertainty regarding the task to be performed, a subordinate will be unwilling to respond to a participative style.
4. Past experience; has the Practice historically been used to shared decision making?
5. Cultural factors; younger people, who are now more highly educated than entrants to the construction professions in previous generations, tend to demand more involvement. Also, as the diversity of operations of a Practice increases especially into the 'meta-professions', such as project management, facilities management and other areas requiring more creative skills, the different style of person employed will tend to require more responsibility for their own work.

The task or job to be done

The nature of the task to be done can often determine what is the most effective leadership style for that situation. Does the task require initiative in problem-solving (cerebral skills) or is it predominantly mechanistic? In the first case a consultative style is generally more effective whereas a directive style is more suitable for the second case.

The time available for the job can certainly affect the most appropriate leadership style as a 'rush' job (which most jobs in professional offices seem to be) will often be better led by a directive style.

However, as mentioned before, a creative or problem-solving task will often require more time and will be of a more complex nature than a mechanistic task such as producing standard documentation. This will require a greater degree of intellectual freedom from team members and make structured leadership inappropriate. However, if the task is considered relatively unimportant then a directive style may well become suitable to prevent 'Parkinson's law' operating, where work tends to expand to fill the time and resources available for it.

Analysis of the 'fit'

For any given leadership situation in a diverse professional office, these three factors can be plotted very roughly along the scale shown in Figure 9.3. The objective of the exercise is to remove bias of trait and natural style from leadership problems and substitute objective analysis of what

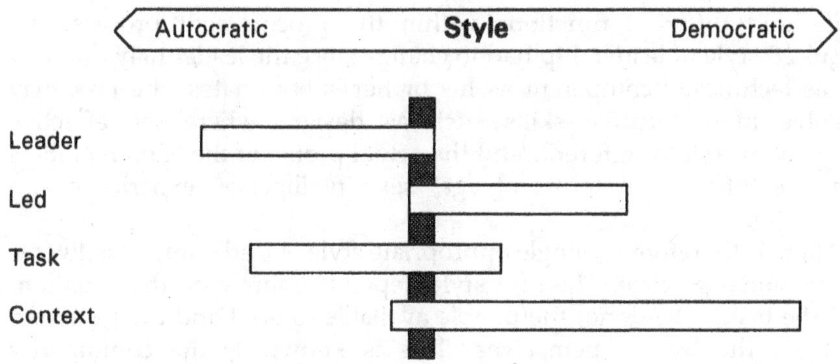

Fig. 9.3 The Contingency Theory of leadership styles.

style is most appropriate under the circumstances. In many cases there will not be a perfect fit but a rough idea should be provided as to how best to achieve the task given the people and resources available.

9.6 CONCLUSIONS

Professional organizations are undergoing major structural changes. In order to compete in the more sophisticated construction markets in the 1990s and beyond, where traditional demarcation does not exist and foreign competition is increasingly intense, they are being forced to diversify their operations in order to maintain market share. There will always be firms still specializing in the 'traditional' service but this will inevitably become a niche strategy, in the same way that small traditional grocer's shops have had to adapt in the face of Sainsbury's, Tesco, etc.

Construction procurement is also changing along similar lines, with the 'one stop shop' (Design/Build and Package Deals) becoming more popular especially with the industrial and commercial clients.

Both of these changes to the industry and profession have had their effects on the way that professional practices are managed. Traditionally, professional practices were not considered as businesses, therefore management was never really considered as a separate function, as we mentioned in the Introduction to Part Two. Leadership was therefore left to the particular trait or natural style of the partner taking responsibility for the task or project in hand. This has been shown to be reasonably effective when the leader is technically competent in the task and the task is predominantly mechanistic as many of the functions in the traditional procurement system have tended to become.

However, as practices diversified to encompass different technical skills and different functions within the procurement process, the required style of leadership had to change since the leader may not now be as technically competent as his or her subordinates, the task may involve more creative skills such as design, where the effective leadership style is different, and the actual profile of the team members may be different in terms of age, sex, intelligence, experience and motivation.

There is therefore no single appropriate style of leadership in a diverse professional practice. The best style depends entirely on the situation, i.e. the task to be done, the people available to do it and the preferred style of the leader themselves. This is known as the contingency approach.

9.7 REFERENCES AND FURTHER READING

Adair, J. (1988) *Effective Leadership*, Pan Books Ltd, London.

Crosher & James, Chartered Quantity Surveyors (undated). *Which Construction Procurement method?*

Franks, J. (1984) *Building Procurement Systems*, Chartered Institute of Building, Englemere, Berks, UK.

Greenhalgh, B. (1990) A general classification for the procurement of construction using the basis of functions and responsibilities. CIB Symposium W92, Zagreb, Yugoslavia, September, 1990

Handy, C.B. (1985) *Understanding Organizations*, Penguin Business Books, London.

Mathurin, C. (1990) A study of responsibilities, guarantees and insurances in the construction industry with a view to harmonization at community level. Commission of the European Community, Brussels.

Open University (1989) Course documentation for module B600. 'The Capable Manager'. The Open University, Milton Keynes.

PART THREE

Strategic Management and Marketing of Professional Services

Introduction to Part Three: Managing the unfamiliar, uncertain and unpredictable

Professor Peter Lansley; Dean, Faculty of Urban and Regional Studies, University of Reading

INTRODUCTION

After a brief consideration of the different approaches to the strategy formulation process this introduction considers two popular vehicles for improving business performance – Quality Management and Information Technology. It argues that in order for any performance-enhancing initiative to have a lasting impact senior management must have a sound grounding in both the philosophy and practice of the chosen approach.

The paper considers the changing nature of the markets for the services of land, property and construction professionals and suggests that viewing these in terms of constituent market sectors provides a good starting point for formulating strategies and developing marketing policies.

Finally, we consider some of the market factors which are shaping and reshaping the identity of the professions. It suggests that just as Project Management and Facilities Management have emerged as new professions which embrace many of the functions of existing professions so a further new profession may be ready to emerge. The response of the existing professions to the market forces which are demanding this new approach will have a critical influence on the status of the professions over the next decade.

It is a very great pleasure to introduce this Part, especially as the theme is so close to my personal and professional interests. Indeed, the theme, strategic management and the marketing of professional services, is one which tempts me to reflect on 25 years as a lecturer, consultant and researcher. Throughout that period I have been in search of the answer to that elusive question – 'What are the key ingredients of corporate success in the land, property and construction professions?'

This is a question which elicits all sorts of answers, not least from senior partners and executives; answers which provide vital clues to how those individuals manage their firms strategically.

Sometimes, perhaps too often, there is a view that as we are in an industry which is not dependent on heavy investment in plant, equipment or hardware, the key ingredient is the people in it, their skills and their enthusiasm for their work. In part that is true, our businesses are only as good as the people in them; indeed, it is they who embody the technology of our industry.

Surprisingly few mention having a clear view of the market and exploiting that view. There is little reference to strategic management and marketing. Sometimes this is because an understanding of the market is taken as read. At other times so is the issue of how to market professional services. In short there is a paradox. If ever there was an industry which is greatly affected by the vagaries of the market it is ours, yet much of the time we seem to deny that there is anything special about that market.

Understanding the paradox is easy. To a very great extent many in the industry are so used to living with the vagaries of the market that as a means of survival they have adopted approaches to business which accommodate those vagaries. Often, however, these have led to a minimalist approach to developing our profession and in meeting the needs of society. The approach has been that of the market trader or merchant so ably described by Michael Ball in his book, *Rebuilding Construction*.

Twenty-five years ago the world was a much simpler place for professional practices and for construction firms. Understanding what made them tick and what contributed to their success and failure was straightforward. Then it was possible to discern clear relationships between, on the one hand, the performance of a firm and, on the other, its internal management and organization. Taking these into account along with a knowledge of the task of the firm – in other words, the type of work it was carrying out and the market it was operating in – it was possible to provide some very clear advice about how the firm could improve its performance and go about managing its affairs strategically.

By the mid-1980s the world had become a more complex place. The heady days of post-war growth were in the distance, survival was more difficult and, as a result, there was a ready market for hints and tips on how to improve business performance as well as for new and revealing insights on strategic management. Management touchstones were offered in some extremely well-written books such as *In Search of Excellence*. In those cases where the books were really understood there were benefits to be gained. In the course of little under a generation the previous somewhat deterministic and instrumental approach to viewing performance was replaced by a dynamic and holistic approach.

Today, the world is yet a different place. Some of the lessons of the past remain valid but there are new challenges requiring different answers. After the quasi-metaphysical advice given in the 1980s the quest for improved business performance has turned almost full circle to systems of management. However, approaches to management based on, say, Total Quality Management (TQM) and Business Process Re-engineering also have a philosophical base but this has not always been appreciated. The attraction of the systems aspects has been too attractive, potential users rarely delve under the surface. Perhaps this is best illustrated by two examples, Quality Management and Computers and IT.

QUALITY MANAGEMENT

One of the most popular current strategic approaches to seeking improvement to managerial effectiveness, particularly in the marketing of professional services, is through quality improvement programmes. Through these programmes firms can gain a competitive edge in those markets where clients understand and value quality. This involves an increasing number of clients. But, despite a proliferation of Total Quality (TQ) programmes, their success has been modest. While many claims have been made, at best the programmes have led to incremental improvements in the performance of firms and in the satisfaction of their clients; at worst they have made it more difficult for firms to increase their competitiveness.

TQ programmes have often failed because in many cases they have been internally focused. They have lacked a clear link to clients or to business performance. Rigid predetermined TQ programmes imposed on an organization do not work. Those firms which have benefited most have done so from TQ developing organically, building on a sound understanding of the philosophy of TQM. Yet firms ignore TQ at their peril. They need to discover how to achieve higher levels of performance, sustained by continuous methodical improvement.

Our knowledge of what can be achieved in terms of quality and client satisfaction suggests that TQ programmes have a natural place in our professions. However, the success of such programmes will depend on those factors which underlie the success of any other performance-enhancing scheme, the vision and commitment of senior management – in this case an overriding concern for quality.

It is not viable for senior managers to take an arm's length view of performance-improvement programmes. For example, reading about the benefits of TQ is not sufficient. The whole rationale of TQ demands working through the theory in detail and applying it to practice, with the top level managers involved as well as the lowest level. Starting at the top, the chief executive has to improve the quality of the work of the board, before being able to move to other levels, sharing skills and changing attitudes.

One issue is that when looking at the people aspects of client relationships a wide range of interpersonal, communication and information skills are involved. It is not just a matter of design, engineering, technical and contractual knowledge. Knowing how to undertake certain tasks has to be matched by an understanding of why it is necessary to undertake them. For example, senior managers need to know why they have to treat employees in a thoughtful and polite way. Only then will they treat clients in the same enlightened manner.

COMPUTER AND IT INITIATIVES

Another vehicle for managerial and organizational development is the implementation of computers and IT. This has immense implications which can be appreciated fully only by sensitive business planning processes. New IT provides for greater decentralization and geographical dispersion of work and it calls into question traditional organizational boundaries, for example between a firm and other members of the building and property team. With the ability to access information which is continuously updated the role of the professional and of relationships with clients changes dramatically. Contact through the hierarchy becomes more frequent, decision making becomes more public and explicit. Professionals become more exposed to scrutiny. While the change in cost structures brought about by IT clearly influences the competitiveness of a firm, much depends on whether that firm views IT as enhancing existing functions or as providing completely new ways of thinking about business. This will depend on the extent to which senior management has a working and all-embracing understanding of IT rather than a passing familiarity with its potential.

Unfortunately, for many construction professionals, familiarity is low. A major skills gap exists and even where strategic planning does take place consideration of IT does not figure strongly in the process. Many are unconvinced about the need for IT or because the process is not led from the top are overwhelmed by the politics of organizational change.

THE CHANGING TASK

So far I have commented on a number of familiar developments without saying why these are taking place or at least should be taking place. In fact I have been like the senior managers who forget about the market when offering their key ingredients to success in business.

Industries, firms, individuals do not, as a matter of course, change the way they operate except in response to changes in their environment. Quite simply, presently the professions are being turned upside down not through choice but because circumstances require change. The markets in which firms operate have shifted, the services required by those markets are different. In short, the overall task of the firms has been transformed.

BUYER'S MARKET

First, we live in a buyer's market. There is overcapacity in most sectors of the industry. Clients are more knowledgeable and demanding. They call the tune. For any service which a client may require there are many more alternative types of supplier than in the past. For example, contractors who used to stick to just the construction phase of a project are now involved in the front end of projects, carrying out feasibility studies and putting together financial packages. Also, they are involved in the rear end, managing the facilities.

Recently I came across a firm which for many years had developed and run hotels. Now it is acting as a consultant to a university on the refurbishment and management of its halls of residence. They have managed these projects through every conceivable phase. Their client is exceedingly content with their performance, so much so that the agent it had used for many years has been displaced.

On another recent occasion I came across a team which was project managing a package-deal hospital. The team had more people with medical qualifications than staff with construction qualifications. That project characterizes today's condition in the industry, that the actors in the construction process are but pieces of a Lego kit. They can be selected and put together in strange and unexpected ways.

OPEN MARKET

Secondly, we live in an open market where new players are entering at a bewildering speed and are playing new games – games which many clients find very attractive. Some 25 years ago, chartered accountants audited a company's accounts, management consultants gave advice, architects designed, builders built and surveyors held the client's hand. Now, with the Lego kit approach, every thing is up-for-grabs. Competition comes not just from other types of professional firm but from organizations with a very different background. Some new entrants are bigger. They have much more muscle than established firms in the industry. For example, the water companies have rapidly expanded their engineering consultancy operations into organizations which seem quite formidable.

Others have entered land, property and construction through a very different way of viewing the needs of clients, for example, electronics and computer firms are involved in facilities management because of their expertise in intelligent monitoring and automated control. Privatization has encouraged the growth of the new forms of competition but so has a return to core businesses in the private sector, although what is regarded as a core business has changed.

GLOBAL MARKET

Thirdly, the open market is a global market. International players appear in many guises, as overseas firms, as owners of well-established UK firms and as joint-venture partners. Each combination adds to the variety through providing yet another blend of skills and expertise.

Understanding the factors which drive markets and which encourage new forms of competition is an essential ingredient to effective strategic planning and marketing. However, as I suggested earlier, often this is neglected, partly through over-familiarity but also because of an aversion to notions and ideas taken from the mainstream of management practice, especially manufacturing industry.

It is not many years since the notion of viewing the construction market as a series of sectors each with a definite life-cycle was met with scorn from the professions. The idea that firms could pick and choose to be in those sectors which were bucking the economic trend was seen as little more than encouraging even more recklessness in what is already a high-risk business.

Yet sectors do follow definite life-cycles and the benefits of being on the rising trend of a new life-cycle have been made clear by many companies. Consideration of how practices and firms have responded to the markets for retirement homes, leisure centres and various forms of social housing provide ample illustration of the value of having sufficient insight to be ahead of the pack. But market sectors come in different shapes and forms, not just types of building. Viewing a market as comprising a number of geographical sectors can be very revealing. Even more powerful is when markets are viewed in terms of procurement method. More than any other approach these define the key players and their relationships with each other. If ever you want a good illustration of the life-cycle of a sector, of fashion, it is in this area. Consider, for example, the growth of management contracting and its demise, the development of construction management, design-and-build and, now, partnering.

When I look at a firm's market sectors in this way I like to undertake a traditional form of portfolio analysis, following well-established principles for identifying the sectors which are the generators of profit and cash, those that need investment, others that need more attention in order for the firm to benefit, and those which are best left alone. There are many techniques for helping with such an analysis but the most celebrated is the Boston Matrix (illustrated below).

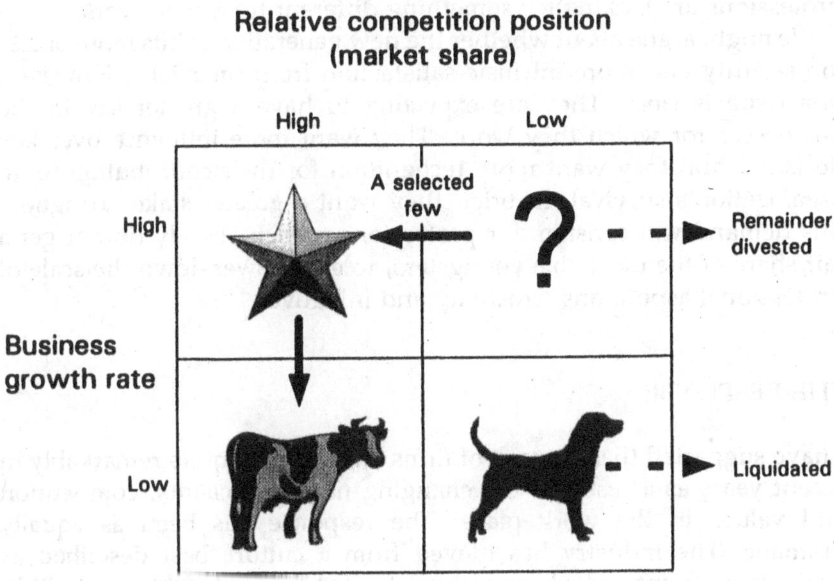

The Boston Matrix.

One of the nice surprises from using this approach is that it works just as well when applied to a business split into its geographic territories, type of service and procurement method as it does when applied at the larger scale to whole businesses.

Increasingly, however, I have difficulty. My analyses go against the grain. They suggest that firms should turn their back on their core businesses, where the competition is great and the returns are low, and should favour relatively new businesses which might be regarded as peripheral to the main interests of the firm. The dilemma is what should be the core businesses for the future.

VALUE SYSTEMS

Markets change – clients change – so do their expectations and demands. The competition changes as well. At the same time there is a revolution taking place within our own firms. The fourth change is not a revolution which many of us might have wished for, but it is there – the impact of the changing values of society and the expectations of the work force. Compared with previous generations the new recruits to our professions are looking for something different from their work.

We might argue about whether the new generation wants more or less job security and more intrinsic satisfaction from their jobs. However, one issue is clear. They are expecting to have a greater say in the businesses for which they work. They want more influence over key decisions and they want more recognition for their contribution to an organization's survival. In brief, they want a greater stake. To ignore this demand will consign our professions, which already do not get a fair share of the most able youngsters, to even lower down the scale of professional aspirations, creativity and initiative.

THE RESPONSE

I have suggested that the task of firms has changed quite remarkably in recent years as a result of the changing nature of clients, competition and values in the work place. The response has been as equally dramatic. The industry has moved from a culture best described as production-orientated where each profession was content to stay within its own well-defined boundaries to one where it was more conscious of the roles of the respective professions and sensitive to the services which it offered. It has moved on from there to a more federated culture, one where through force of circumstance it has had to give a great deal of discretion to its clients and has then had to work closely with them, redefining the services required and hence the roles of the professions.

The most vivid demonstrations of progress have been the development of multi-discipline firms, the move away from partnerships to corporate structures, the establishment of a new profession, a meta-profession, of project management and the emergence of another, that of facilities management. Both seek to redefine professionalism in terms of the services which clients require, that is in terms of the market, rather than in terms of the functions which contribute to the required service.

I do not know which new profession will follow but I am convinced that if the land, property and construction professions are to be involved then they will have to recognize the very considerable criticisms which clients make of what is currently offered. Those criticisms can be summed up for construction firms as the lack of a Total Service. For professional practices this translates into an unexploited opportunity best envisaged as Consultancy Advice Across the Board. Also, I know that just as those firms which have made the grade in project management and in facilities management have done so through getting to know the environment, the markets and their own capabilities, those which will lead the way in developing the next new profession will also have to think and act strategically.

Consultancy Across The Board will develop in many shapes and forms depending on the starting point of any particular firm and it will take time to develop an effective approach. Clearly, the first step is the most important and most difficult, but I am confident that firms large and small can do this. In the last few weeks I have spoken with three such firms, all small. The first was a solo architect which now offers legal advice to its clients and to those other businesses with which it works, especially on the handling of arbitration. The second was a small traditional contractor which had entered into design-and-build. The third was a surveying firm which had created a niche in making chief executives of largish local private sector and public sector organizations more aware of the value and potential of the property assets of their firms and then following up their resulting interest.

The link between these three examples is not just that the firms were following the growing market for these services, but rather that because they understood what was driving the market they reorganized their operations, they invested in training and development and, then, through following the market, each in its own way was successful.

This brings me back to the theme of this section – Strategic Planning and Marketing of Professional Services; that is, managing the external environment. But I am also aware of the theme for Part Four – Maintaining Quality Services and Professional Ethics – because I believe that these contribute the basis for achieving effective planning and marketing.

For the final part of this introduction, I would like to pursue a rather different theme. This is that if our professions are to survive and not be absorbed into some bland and unexciting form of general managerialism, then strategic planning, marketing and quality have to become second nature, built on sound professional ethics. A second nature, however, is not developed easily. It must be achieved through better education and training. Presently much of our professional education is too insular and too narrow to be of value to the clients we wish to serve. And because of the overwhelming vocational nature of many education schemes an inadequate share of very able people is attracted to our professions.

It might be expected for me to say that the vocational knot which constrains so much education has to be untied, and that management and business education has to play a much greater part. However, I would go further. I would suggest that technology education is wholly inadequate for preparing many land, property and construction professionals for the task ahead. Technology education has to become more generic and transferable – understanding the technologies employed by clients is as important as understanding the technology of construction. By the same token, management and business education has to facilitate the rapid assimilation of the many different organizational, managerial, economic and legal processes which influence the businesses of clients. If this cannot be achieved then we will see the continual replacement of existing professionals by those from outside to which I referred earlier.

Perhaps I am too pessimistic. Earlier I mentioned the emergence of two meta-professions, project management and facilities management. Although these have largely cut across the established professions they coexist with them. Perhaps the future will be one where there is a two-tier professional structure, which in many ways will be analogous to that which used to exist, of professionals and technicians. The danger is, however, that our professions will be in the second tier.

CONCLUSION

In setting the scene for the themes Strategic Planning and Marketing of Professional Services, I have deliberately considered the 'why' rather than the 'what' and the 'how'. In so doing I may have implied a task for the development of the professions which is not only daunting but impossible. However, today's professionals have at their disposal a range of tools, techniques and technologies which are commensurate with that challenge.

I started by alluding to the fashions of the 1980s for simple touchstones for achieving management success. While the blind belief in such an approach to the improvement of performance has been discredited, much of what was written at that time was sound. Perhaps the most sensible was the simple message – that success comes from trying new approaches, from experimenting, from taking action.

If ever there was a time when land, property and construction professionals needed to follow this advice it is now. As a result of this conference or perhaps in spite of it, if every one here should resolve to do something different in the way they plan and market their businesses and to keep taking new actions then I am sure the professions will have an exciting and rewarding future.

REFERENCES

Ball, M. (1988) *Rebuilding Construction*, Routledge, London.

Dixon, T. (1994) *Skills Training and Education for the Surveying Profession: Requirements for the 1990s*, CEM, Reading.

Gibson, V., Jones, K., Robinson, G. and Smart, J. (1995) *The Chartered Surveyor as Management Consultant: An Emerging Market*, Royal Institution of Chartered Surveyors, London.

Hedley, B. (1977) Strategy and the Business Portfolio, *Long Range Planning*, February.

Hillebrandt, P., Cannon, J. and Lansley, P. (1995) *The Construction Company in and out of Recession*, Macmillan, Basingstoke.

Lansley, P. (1987) Corporate Strategy and Survival in the UK Construction Industry, *Construction Management and Economics*, **5**(2).

Lansley, P., Sadler, P. and Webb, T.(1992) Organization Structure, Management Style and Company Performance, *Omega*, Vol. **2**, No. **4**.

Peters, T. and Waterman, R. (1982) *n Search of Excellence*, Harper and Row, New York.

CHAPTER 10

Creating a sustainable competitive advantage and managing prequalification team presentations

10.1 INTRODUCTION

This chapter considers the necessity for property and construction companies to be able to create a sustainable competitive advantage in order to assist in their long-term survival, and discusses the management of pre-qualification team presentations in contractual services. Too often companies put short-term gain before long-term stability, with the effect of undermining the ability of the organization to survive the cyclical nature of demand for the products of the industry.

In it, we draw upon recognized management theory and apply it to the property and construction sectors of the economy. Lessons are established which may be learned from traditional management theory and which can be utilized to assist property and construction companies to create a sustainable competitive advantage over their competitors.

The property and construction sector of the UK economy is characterized by a large number of small to medium-sized enterprises. For example, in the UK construction sector in 1992 over 94% of all firms fell into the size category 'up to seven employees'. These sectors of the economy, in common with many other sectors, are also characterized by a high number of company failures.

In attempting to prevent company failure there is a need to create a sustainable competitive advantage for the company within their specialist area of expertise, locality or region. The emphasis must be on sustainable, because as one company creates a competitive advantage, its competitors will imitate or recreate this characteristic and the competitive edge will be lost. Once the competitive edge is lost, the company must endeavour to create another distinguishing feature in their service or product and so the cycle begins again.

10.2 COMPETITIVE ADVANTAGE

Michael Porter sets out a number of generic strategies which may be followed when seeking to create advantage for a company. These are mainly concerned with the adoption of an approach which concentrates on cost leadership or differentiation. Cost leadership is characterized by the company being the low cost producer in its market segment. With increasing emphasis today on quality and the insistence of clients on quality in terms of service or product, I would suggest that it would generally be inappropriate for property and construction companies to adopt this strategic model. This is because this strategic model is often characterized by a concentration on cost to the detriment of quality. Everything is undertaken to drive down the cost of the product or service so as to ensure that it is consistently seen as the **least cost** option. The necessary questions when considering a cost-based strategy are as follows:

- Can cost be reduced?
- Can value be maintained?

If the cost of the product or service to the client can be reduced without adversely affecting the utility of the commodity, it may be possible to pass the cost saving on to the prospective client, to increase profitability or a combination of both.

However, differentiation as a strategic model, permits companies to distinguish themselves from their competitors in one of two ways:

1. **Uniqueness**: this may be in terms of product or service. It is possible by careful analysis of client need to ensure that the product or service being offered not just meets the minimum expectations of the client but that it exceeds expectations.
2. **Premium pricing**: if it is possible to create added value to the client in the product or service being offered, Porter maintains this may lead to the possibility of charging a premium price.

Considering uniqueness first, the property and construction industry must ask itself if the products or services which it offers can be differentiated from the general competition. It is often said that clients of the property and construction industry are concerned with three things – cost, time and quality. I will call these three elements **competitive factors.** It therefore follows that property and construction companies should endeavour to differentiate themselves in as many of these competitive factors as is possible. It is the critical combination of these factors to any one particular client which is likely to give a company a sustainable competitive advantage.

What is needed is a knowledge of the individual client which permits the company offering the product or service to know which combination of the three competitive factors is crucial to their potential customer's business. As each client is distinct in terms of their needs, the balance of these competitive factors will be dynamic and will need to be ascertained in each bidding situation in which the company finds itself. It is not sufficient to merely consider these three factors but an analysis of client need should be undertaken to ensure that an optimum balance of the competitive factors is achieved.

10.3 FAILURE STRATEGIES

It may seem unnecessary to consider failure strategies as no company would knowingly wish to employ a strategy which leads to failure. However, some companies through lack of knowledge do adopt a strategy which is likely to lead to failure. Those strategies which are doomed to failure include:

1. Increasing price while maintaining the value added to the client. This is often adopted by monopolies and does appear to work when the switching costs to the client are too great to warrant going to an alternative supplier of the product or service. For organizations where there is no clear advantage to the client to remain with them, this may achieve a short-term improvement in profitability but is almost certain to lead to failure in the long-term.
2. Increasing price while reducing the value added to the client. All too often cost increases are introduced to reduce budget deficits, with little regard as to whether or not the client sees the product or service as being value for money. If the cost increase outstrips the worth or value to the client, they are likely to seek better value elsewhere.
3. Maintaining price while reducing value added to the client. As with the previous strategy if the client considers the cost to be too high in relation to the worth of the product or service, it is likely they will switch to an alternative.

These strategies, by their concentration on cost and not on value will almost certainly lead to failure. Value analysis offers the potential for the supplier of products and services to the property and consfruction industry to undertake an analysis which has the potential of reducing cost while maintaining the value to the client. This cost saving may enable the company to offer part of this benefit to the client, thus ensuring competitiveness.

10.4 SUSTAINABLE COMPETITIVE ADVANTAGE

Sustainable competitive advantage can be achieved by:

- identifying potential client needs and values;
- establishing which generic strategy route is most suitable;
- operationalizing this strategy;
- achieving cost efficiencies/reductions, where appropriate; and
- checking that strategic directions and methods of organization are in line with the generic strategy.

Morrison and Lee have tried to identify the characteristics of successful enterprises. They state that successful businesses:

- identify key success factors – things which are critical to the business if they are to remain competitive;
- measure and analyse competitive advantage;
- anticipate their competitors' responses;
- exploit degrees of freedom; and
- invest in competitive advantage.

Perkowski looks at the future of the engineering and construction industries and states the following are characteristics of successful companies. In his opinion they:

- understand change;
- adopt a systems approach;
- accept mistakes happen and reward sensible risk-taking; and
- provide innovative services.

In the opinion of Ohmae, successful enterprises:

- identify key success factors;
- exploit relative superiority – use the advantage while it remains; this recognizes that any strategic advantage is likely to be short-term until competitors can catch up;
- attempt to change key success factors – even when the key success factors are not in favour of the company it may be possible to exploit them to create an advantage; and
- innovate – as strategy is a dynamic process, companies must be prepared to take calculated risks in being innovative.

Key success factors are therefore identified as 'anything which helps a business achieve or maintain its competitive advantage'. Prahalad and Hamel talk about the need for companies to develop core competencies, while Stalk and colleagues use the term 'strategic capabilities'. Pampin focuses on what are called 'strategic excellence factors'. This concept is identified as focusing on issues which matter to the customer (e.g. quality, image, advertising and branding).

10.5 PROJECT PARTNERING

The construction sector of the economy particularly is characterized by a client base which is one-off. This creates an inability to form a long-term relationship with clients for the benefit of both client and contractor. Baden Hellard, in his book, *Project Partnering – Principles and Practice*, sets out a concept which seeks to establish meaningful relationships between client and project team, even for the one-off project. These relationships are based upon trust and seek to establish

> . . . an environment where trust and teamwork prevent disputes, foster a cooperative bond to everyone's benefit, and facilitate the completion of a successful project.

The concept of project partnering will figure increasingly in the relationships within the property and construction industry over the next decade and do provide an opportunity for businesses to differentiate themselves from their competitors in terms of creating a caring relationship for the needs of the client.

Baden Hellard states also that project partnering brings with it certain benefits to the client, the design team, contractors and sub-contractors. These can be summarized as follows:

1. Reduced exposure to litigation.
2. Lower risk of overruns in terms of cost and time.
3. Better quality product.
4. Reduced overhead costs.
5. Greater profit potential.
6. Involvement in problem identification and solution.
7. Enhanced opportunity for repeat business.

These benefits are exactly what most companies would wish to achieve by undertaking a strategic analysis of their business when attempting to create a sustainable competitive advantage. Perhaps part of the answer to creating competitive advantage must therefore lie in project partnering.

If these benefits can be realized for an individual project, they will be to the ultimate advantage of the industry as a whole. In Baden Hellard's words, 'Partnering has the potential to change our industry, one project at a time'.

10.4.1 Conclusions

Businesses must concentrate their strategic efforts on those issues which matter to the customer and should seek to ensure that they adopt a strategy which capitalizes on those aspects of the competitive factors which, in the eyes of the client, make their product or service distinct from their competitors. The client should be the beginning and end of strategy within the property and construction industry.

Strategies which concentrate purely on cost may fail to consider the value to the client and may thus lead to failure. In an industry dominated by small to medium-sized firms, there are many organizations offering similar products and services to the industry. This should lead companies to seek to differentiate themselves from their competition in terms of value to the client. This involves a detailed analysis of client needs and a matching of the chosen strategy to the particular needs of the client or group of clients. Where companies have a long-term relationship with a particular client, this may be relatively easy. However, if most clients of the property and construction industry are one-off, this process is much more difficult. The processes adopted traditionally within the industry often use experts whose function is to act as agents on behalf of the client. In these circumstances it is their responsibility to convey to the product or service provider a detailed brief which sets out those matters which are critical to the particular client.

Project partnering may hold the key for small to medium-sized companies within the property and construction industry in enabling them to adopt a style of strategic management which inherently promotes competitive advantage. This would not be based on the traditional model of confrontation, but rather on those aspects of the competitive factors which are of importance to the client. This should ultimately be of benefit to the property and construction industry as a whole.

10.6 THE MANAGEMENT OF PRE-QUALIFICATION TEAM PRESENTATIONS IN CONTRACTUAL SERVICES

More demanding and knowledgeable clients and increased competition has prompted both contractors and professional practices to give more consideration to the marketing function within their organizations. Pre-qualification procedures are now used ever more widely by clients in evaluating contractual services, which allows for competition based on both a reputation for quality service and on price. Pre-qualification presentations enable clients to make a more informed choice in their procurement process and therefore adds value to both the service being provided and also to the finished building.

The principal recommendations in the development of an effective pre-qualification strategy are that contractors and professional practices should be more proactive in establishing clients' needs and expectations. Testimony from previous satisfied clients should be used to illustrate the relevant track record of the firm wishing to be considered. There is therefore a need for a shift in the culture of organizations in construction to achieving the total satisfaction of clients.

Our purpose here is to identify the main issues in the management and organization of pre-qualification interview and team presentations in contractual services in construction. Despite its significance in commercial terms, this is an area which has received little attention from academics and other writers. Other than a number of articles in the construction press and professional journals, this vital process in the competitive strategies of contractors and professional practices in construction is underdeveloped.

Previous research into the marketing and promotion strategies of major UK construction organizations, and which investigated the expectations of major procuring client organizations across a number of sectors (retail, manufacturing, the utilities i.e. gas, water, electricity, etc.) concluded that the marketing and promotional function in the majority of construction organizations is underdeveloped and not contributing towards providing a competitive advantage. There was a lack of satisfaction from clients and their consultant advisers i.e. architects, quantity surveyors and engineers, with the promotional efforts of contractors and consultants offering their various services.

Additionally, organizations should be developing a more planned approach to the marketing and promotion strategy, giving attention to the gathering of information on clients and competitors, and providing more differentiating messages concerning their management services.

Given that clients were found to be seeking a long-term partnership between themselves, consultants and contractors, emphasis on the development of better communications, negotiation and customer contact skills was seen to be required.

The empirical evidence suggests that there are clear differences between what organizations are promoting, the messages they are conveying, and what clients and their advisers expect from marketing and promotional activities.

Although there is a need for organizations to use a variety of marketing techniques, i.e. press relations, publicity brochures, media advertising and project site-based promotion, the point of convergence for obtaining a competitive differential advantage in contractual services is through direct selling and the process associated with pre-qualification procedures.

We will now attempt to establish the nature and content of pre-qualification procedures and identify important issues in the development of a strategy for providing more effective and competitive presentations of contractual and professional services.

This section is based on a literature review, combined with the results of a number of exploratory interviews conducted with major client organizations, contractors and architectural, quantity surveying and engineering consultants.

10.7 PRE-QUALIFICATION PROCEDURES

Pre-qualification procedures largely depend on the client and the design team advisers but have generally been introduced into the procurement of construction work due to the need to be able to judge between competing firms. The selection and appointment procedures in the pre-tender process offer contractors an opportunity to differentiate themselves from the competition through their reputation, record for successful completion of similar work and corporate brand image.

Pre-qualification usually entails compilation of a list of contractors or consultants, use of telephone and mailed questionnaires, and requests for contractors' accounts and background information. The outcome of this process is a short list of contractors for interview or to make presentations.

Initial telephone interviews may be used to establish the following basic information on the firm:

- turnover
- value of contracts secured to date
- whether the firm is willing to submit a tender
- how long it will take to complete the tender

These initial contacts would be followed by selection questionnaires which would cover considerably more detailed questions on annual turnover over a number of years and the value and details of projects of similar size and complexity.

Direct face-to-face interviews may be used to gain a detailed description of the overall project, the construction programme and an explanation of the terms and conditions of the contract, including responsibilities of the parties. The content of the interview would include organization of the project; site administration and project team, setting out and dimension control, material handling and control, contractors supervision of on-site responsibilities, site safety, labour relations and quality management.

Interviews would be concluded with decisions over the period of tendering, setting of mid-tender interviews, and actions required and date deadlines.

The following section will consider the responses to an exploratory interview survey of client organizations which aimed to establish their expectations of pre-qualification interviews and presentations.

10.7.1 Client expectations

Pre-qualification allows clients to assess contractors and consultants on their expertise, experience and ability to carry out the proposed project before tender price or fee submitted. Increasingly, contractors are expected to demonstrate an understanding of the client's business; its products and markets.

The qualifying factors for inclusion on to select tender lists as established in this survey include the following:

- financial standing
- reputation for carrying out similar work at the appropriate level of quality
- quality of the project team, managerial and communications skills, abilities to solve contractual problems
- approaches to the work with minimal disruption to the client's business; staff and customers

Clients have indicated that contractors may gain an advantage if they provide detailed and credible references from previous satisfied clients, can demonstrate their interest in finding out as much as possible about the client and proposed project, and provide a single point of contact within the organization from first approaches in the pre-qualification period, throughout the project and into the maintenance stages.

Clients expect that those contractors or consultants who are invited to submit proposals are positive and professional in their approach. They need to provide a full team composed of project and site managers, including some who would be responsible for the proposed project, and others with extensive contractual experience. It is important that someone – possibly at director or partner level – would be present who would be accountable for the firm and be able to draw on his or her expertise in resolving any contractual difficulties foreseen.

10.7.2 Contractors and consultants – current practice

Contractors and consultants interviewed in the preparation of this paper indicated that any marketing function within contractors is mainly responsible for producing brochures and other publicity material which support the overall marketing efforts of the firm. There was an overall consensus that marketing was important in establishing contacts and conducting research into the market, but that its role in terms of pre-qualification presentations and interviews was very limited. The client was interested in seeing those senior managers within the contractor or consultant organization who would be responsible for the proposed project.

A portfolio of printed material is developed by many firms and practices to be used in conjunction with presentations. These submission documents may include professionally produced brochures and project sheets which communicate:

- the organization profile;
- financial standing, possibly through bankers references;
- recent relevant experience; and
- CVs of relevant project management.

Issues in the production of the material highlighted by contractors and consultants were the need for flexibility to be able to tailor packages to individual clients and current projects, the cost and time required to put together documentation and material becoming out of date.

A number of those interviewed stressed that generally improvements could be made to the quality and design or material, but this would likely make documentation more expensive to produce.

Contractors and consultants acknowledged the need for rehearsal before team presentations and interviews, but in most cases, given that representatives of the firm were highly experienced in pre-qualification procedures, formal training was not seen as necessary.

There was general criticism of client organizations and their representatives. Many contractors commented on the large amount of information required in questionnaires and the fact that the purpose of many questions seemed to be unclear.

There was frustration when it was clear that clients and their representatives had already made their decision as to who was going to be awarded the contract, and when clients failed to provide meaningful feedback on why submissions had been unsuccessful.

10.7.3 Development of an effective strategy

In developing an effective strategy for conducting pre-qualification presentations, it is necessary to identify two possible scenarios. The first is where a company is engaged in actively promoting itself to possible future clients, often through architects, with a view to being invited onto future tender lists. The second is where the company has been invited to respond to the requirements of a specific contract.

The principal objectives in any pre-qualification strategy may be to:

- influence the client and members of the decision-making team when they are in the process of making invitations to tender. This is only possible if the firm or practice takes a proactive stance, and initiates contact with its target markets. This may be through getting potential clients and consultants to current sites, giving impromptu presentations, inviting clients to company seminars, etc.
- develop a clearly differentiated proposal. This may only be achieved

through establishing and analysing client needs and expectations, the offerings of competitors and the organization's own resources capabilities and strengths.
- communicate clearly with the client team and negotiate the best contract.

Effective interviews and presentations depend on the firm or practice gathering as much information as possible on the following:

1. The client organization and decision-making team, and their expectations.
2. The client's products and markets, and problems they may be facing themselves.
3. Competitors' strengths and weaknesses and any relationships they may already have with the client.
4. The strengths and weaknesses of the firm or practice, past or current relationships with the clients, the value of the service to be provided, and experience of previous similar projects.

A firm or practice needs to conduct research which identifies what the client wants, what they already get, what they go to competitors for, and what else they would like. This process allows a firm to compare its services against client criteria and identifies opportunities. These opportunities need to be matched with available resources. This will enable identification of a competitive difference which needs to be promoted through background publicity and brochures, but most importantly through direct contact during interview and presentations to the client.

10.8 IMPLEMENTATION OF AN EFFECTIVE PRE-QUALIFICATION STRATEGY

Implementation of an effective strategy for pre-qualification presentations requires consideration of the following areas: (i) the need to communicate a distinctive and meaningful message to the client and his advisers; (ii) presenting the right team; and (iii) presentation style and contents.

Analysis of the customer, competitors and the firm itself needs to be converted into a differentiating and ultimately winning message. This message needs to satisfy the basic expectations of the client team, that the firm has the appropriate track record, a high-calibre and reliable project team, adequate financial backing and quality management.

In addition, a firm needs a unique selling point (USP) which can only come from analysis of the organization's own strengths and weaknesses relative to the client's expectations. How can the firm or practice exceed the client's expectations? Arguments may need to be presented which very clearly differentiate the firm from its competitors. Disadvantages associated with competitors may need to be emphasized. The firm needs to be able to provide detailed and relevant references from previous satisfied clients or consultants who can provide independent testimony to support proposals.

In larger contractors and consultants some system for sharing client information should be set up across the organization. Regions of the business may have had experience with the same client and this knowledge may provide a competitive advantage.

It would seem essential that a flexible resource of material to support presentation teams should be prepared, kept up to date and coordinated by some form of centralized function within the organization. This may be a single manager within smaller firms, or through departments in larger organizations, responsible for completing questionnaires and developing submission documents to be presented. This function may also be responsible for selecting teams to attend interviews and presentations.

Clients and their advisers are not looking for 'slick' presenters. Presentations need to be led by the project manager who has been allocated on a full-time basis. Clients do not want to see glossy presentations, but hear from people involved from director down to site level. Audio-visual equipment such as videos may enhance the presentations and – if professionally produced – communicate past performance more effectively. However, their use should not in themselves distract from the main message.

Presentation teams will be drawn from different functions of the firm, i.e. contracts managers, quantity surveyors and planners. This requires cross-functional cooperation and communication. In most cases it would seem to be important that a senior manger; a director or partner of the practice should lead the team. Although representatives may be highly experienced technical managers, they may require some training in negotiation and selling skills to enable them to more clearly communicate the flexible problem-solving abilities of the management team and its ability to respond quickly to changes in clients' requirements which may be posed in the interview situation.

It would seem essential for the organization to engage a member of the team who would act as a single point of contact with the client. This person would need client contact skills in addition to being able to answer technical queries throughout the project.

In the event of a firm or practice being unsuccessful, it is essential that the team finds out why. This will enable the organization's management to refine and improve the strategy for future submissions. A constant dialogue with the client and other members of the decision-making team is desirable and promotes a highly client-oriented image.

Personal presentations, combined with brochures and publicity need to be planned and coordinated to communicate a consistent message. Identification of the client's expectations and criteria for selection and competitor's offerings, should be the basis of promotional efforts. The challenge is how to communicate value for money and the anticipated work load, the track record of the firm or practice with particular reference to similar projects, and quality of performance in terms of the work done and also during the maintenance period work.

10.8.1 Conclusions

Many client organizations have developed select tender lists of contractors and consultants with whom they have established working relationships. The challenge for competitors is how to be considered for inclusion on those lists. This paper has identified a number of important issues in the creation of an effective competitive strategy in relation to pre-qualification interviews and presentations.

The need for a planned and coordinated approach demands that contractors and consultants find out as much about the client organization as possible; its history, culture, products, and markets. They need to identify client's expectations, decision-making process and criteria, the members of the decision team and culture of the organization. At the same time it is important to identify the competition, their possible previous relationship with the client and previous work on similar projects. This will enable the firm to communicate a distinctive and persuasive message which is targeted at the specific client and proposed project.

Increasingly, clients are seeking contractors and consultants to provide testimonial evidence from previous satisfied clients. Important considerations are the credibility of referees and the relation of previous jobs to the proposed work. This would seem to be an effective method of communicating value for money, track record, quality of performance, resolution of problems and financial stability.

Contractors expressed concern over the amount of effort required in responding to extensive questionnaires. There appeared to be a consensus that often it was not clear why certain questions were being asked and that it may be desirable for some collaboration between contractors to try and get clients to provide a more standard format.

Regionalization within many medium to large construction organizations may act against obtaining competitive advantage. Opportunities may only be created by sharing 'project value' information relevant to clients. For example, regions or divisions of a firm may have been through similar procedures with the same client organization. The management of presentation teams needs to be supported by cross-functional information and cooperation to gain a clear competitive advantage.

Contractors and consultants need to be more proactive in establishing what clients thought about the information, brochures and presentations that they had submitted. It would appear to be the case that meetings are mainly at the client and his or her adviser's initiative. Contractors would promote a more customer-oriented image if these were to be initiated.

Increasingly, clients are seeking long-term relationships based on a partnership between compatible organizational cultures. This demands greater attention to personal communication, negotiation and customer service skills aimed at providing total customer satisfaction. This requires a cultural shift and active support from the top in orienting the organization to satisfying the client. This should extend to providing a single point of contact within contracting or consultant organizations who would be available throughout the contract and beyond. This would assist in combating possible cultural differences between contractors, consultants and clients and overcome misunderstandings much more quickly.

More detailed research would seem to be required into:

- Clarifying and interpreting the needs of the client.
- Coordinating and streamlining flexible documentation to be used in submissions.
- The selection and coordination of cross-functional teams.
- Sharing client-specific information across organizations.
- Provision of a single point of contact who can provide a problem-solving service to the client throughout the contract and beyond.
- The marketing department, its function and possible role in the coordination of pre-qualification interviews and presentation teams.

10.9 REFERENCES AND FURTHER READING

Baden Hellard, R. (1995) *Project Partnering – Principle and Practice*, Thomas Telford, London.

Bernink, B. (1995) Winning contracts, in *The Commercial Project Manager* (ed. J.R. Turner), McGraw-Hill, London, pp. 325–49.

Betts, M. and Ofori, G. (1992) Strategic planning for competitive advantage in construction. *Construction Management and Economics*, **10**(6), 511–32.

Betts, M., Lim, C., Mathur, K. and Ofori, G. (1991) Strategies for the construction sector in the information technology era. *Construction Management and Economics*, **9**, 509–28.

Building (1991) Adaptable style to charm differing clients, 21 June, p. 38.

Building (1991) Ten ways to get yourself excluded, 21 June, p. 39.

Building (1991) Turning off and turning on the client, 21 June, p. 37.

Building (1993) Are you being served?, 4 June. p. 50.

Building (1993) The eyes have it, 4 June, p. 49.

Code Of Practice For Project Management For Construction And Development (1992) Chartered Institute of Building, pp. 152–6.

Fisher, N. (1991) Marketing. *Chartered Builder*, April, 14–15.
Langford, D. (1992) Book review of international construction project management general theory and practice. *Construction Management and Economics*, **10**(2), 179–81.
Langford, D. and Male, S. (1991) *Strategic Management in Construction*, Gower, Aldershot.
Male, S. P. and Preece, C.N. (1994) Promotional strategies in the marketing of contractual services in UK construction firms. The A.J. Etkin Seminar on Strategic Planning in Construction Firms, Haifa, Israel.
Morrison, R. and Lee, J. (1979) From planning to clearer strategic thinking. *Financial Times*, 27 July.
Ohmae, K. (1982) *The Mind of the Strategist*, McGraw-Hill, London.
Ohmae, K. (1990) *The Borderless World*, Harper Business, New York.
Perkowski, J.C. (1988) Technical trends in the E&C business: the next 10 years. *Journal of Construction Management and Engineering*, **114**(4), 565–76.
Porter, M.E. (1979) How competitive forces shape strategy. *Harvard Business Review*, March–April, 137–46.
Porter, M.E. (1980) *Competitive Strategy*, Free Press, New York.
Porter, M.E. (1985) *Competitive Advantage*, Free Press, New York.
Porter, M.E. (1990) *The Competitive Advantage of Nations*, Free Press, New York.
Porter, M.E. and Millar, V.E. (1985) How information gives you competitive advantage. *Harvard Business Review*, July–August, 149–60.
Prahalad, C.K. and Hamel, G. (1990) The core competence of the corporation. *Harvard Business Review*, May–June, 79–91.
Preece, C.N. (1993) *A Study of the Management of Promotion for Competitive Advantage in UK Construction Firms*, Unpublished PhD Thesis, Heriot-Watt University.
Preece, C.N. (1994) Promoting construction for competitive advantage. *Chartered Builder*, July–August.
Pampin, C. (1987) *The Essence of Corporate Strategy*, Gower, London.
Stalk, G., Evans, P. and Shulman, L.E. (1992) Competing on capabilities: the new rules of corporate strategy; Harvard Business Review, March/April.

CHAPTER 11

Marketing analysis and methods for professional firms

11.1 INTRODUCTION

This chapter looks at the role of marketing in the modern professional service firm and the key tasks of marketing management. How to be successful at marketing is a subject attracting the key attentions of firms of all sizes as well as the professional institutions in the construction and property industries, owing to the greater degree of competition and the relaxation on the rules regarding fees as well as advertising. Other external factors stimulating this focus are greater international competition, rapid technological and social change and shortening product life-cycles. Much is now being written about the marketing of service organizations, especially professional services, and as this matures into an area of marketing in its own right, more techniques will be developed which differentiate it from 'product' marketing, which by now has a very sophisticated set of marketing tools. Professional service organizations have the same problems as manufacturers, they must identify the groups they serve, determine their needs, develop offers to satisfy their needs and communicate with their constituencies. Marketing therefore provides a common discipline for approaching these problems.

A major study by the N.E.D.O. in the early 1980s found that lack of expertise in marketing was the single most disappointing cause of the lack of success of British companies in the international market in the previous two decades. A survey of 21 top US companies found that chief executives in those companies believed that marketing was the most important discipline in their company, and that they see it becoming more important in the future. Whether they come from a marketing background or not, they believe that the development and maintenance of an effective marketing organization is the major requirement for success in the ever-changing environment of today's world economy.

As a discipline, marketing has drawn heavily on economics, especially microeconomics and econometrics, but it is misleading to think of marketing as only a branch of applied economics. Marketing is a catholic discipline, which utilizes findings from several areas; the theory of buyer behaviour draws significantly on psychology, accounting and finance have provided insights for market and product planning procedures, and statistics and psychometrics have provided the tools used by market researchers. Marketing, however has become more than a mixed bag of findings from other disciplines, at its core it is a unique approach to business strategy, which proposes that firms will prosper to the extent that they can understand the needs and wants of clients and potential clients and match their capabilities more effectively to these needs and wants than their competitor firms.

Marketing has borrowed from other disciplines where they can assist in the study of buyers and to facilitating the adaptation of the firm to their desires. Economic theory has proved useful, mainly because it has provided concepts for understanding how the firm should allocate market resources, such concepts have proved especially important in fee calculation and advertising decisions; the real limitation of economics in contributing to marketing has been in the lack of insight it provides for the motivation and choices available to clients.

11.2 MARKETING

Marketing – like economics – is concerned with the way resources are produced and allocated via the exchange process between buyers and sellers in a market. The perspective differs in that, to a marketeer the market means groups of actual or potential buyers and they look at the exchange process primarily from the viewpoint of the seller. Marketing can be defined as the activity of satisfying the needs and wants of buyers through the exchange process in a manner which enables the firm (or other types of organization) to achieve its objectives. The most obvious of such objectives is, of course, profit but market share or growth may be additional ones.

Successful organizations, both in the production and services markets, now approach marketing in a professional and systematic manner, which has led to the recognition of marketing management as a primary business function on a par with finance and production. In his authoritative text, Philip Kotler defines Marketing Management as the

> analysis, planning, implementation and control of programmes designed to build exchanges and relationships with target markets for the purpose of achieving organisation objectives, it relies on a detailed analysis of the needs and wants of target customers as the basis of effective product design, pricing, communication and distribution.

which is as true for practices concerned with the provision of professional services in the land, construction and property industries as it is for any firm dealing with the production of physical products.

11.3 THE MARKETING CONCEPT

It is valuable to distinguish two related ways of thinking about marketing applied to professional services. One is the idea of marketing as a total business philosophy which affects all the different functional areas of the business or practice. The second is a more specific functional notion of marketing which defines the principal tasks of marketing management; those of identifying opportunities, developing policies on fee levels, promotion and advertising, service development and so on. The latter are the tasks which are usually led by the marketing department of a firm, but in professional practices usually takes up the majority of the managerial time of the partners or directors.

The former idea is called the marketing concept which states that the primary function of a business is to satisfy the desires of customers. It emphasizes that failure to do this at least as effectively as competitors, will mean that business cannot survive in a competitive world. Similarly, firms which are outstandingly successful are firms which develop their analysis and planning capabilities to identify or anticipate customer's desires and match them with services more effectively than their rivals. You only need to read the introduction to Part One to see how one firm of consultants in the property industry, W S Atkins, is addressing this issue.

This market orientation is often contrasted with two earlier but still prevalent philosophies of business, one is the production orientation: the view that the product will find the market by itself if it is produced at a low enough cost or with sufficiently advanced technical quality edge; firms holding this philosophy tend to be dominated by engineers or technical experts who pay little attention to the explicit study of buyers and characterizes many firms in the construction industry. Often, such companies eventually come to grief by maintaining services which, while technically sound, turn out to be a poor match with the benefit clients are actually seeking. Other companies are 'sales' oriented, believing that success depends on the effectiveness of the advertising, selling and promotional techniques used to stimulate the awareness and desires of potential clients, rather than on any true differences on the offerings of competitors. The limitations of this philosophy lie in the specious assumption of the gullibility of consumers to sales stimulating devices and the unimportance of the attitudes and behaviour of dissatisfied clients.

The real distinction between the marketing and sales orientations is that selling tries to get the customers to want what the company has; marketing on the other hand tries to get the company to produce what the customer wants. These philosophies that marketing reverses the logic of previous ideas of business were first emphasized by Peter Drucker.

Selling and marketing can therefore be regarded as antithetical rather than synonymous or even complementary; there will always, one can assume, be a need for some selling, but the aim of marketing is to make selling superfluous. The aim of marketing is to know and understand the customer so well that the service will sell itself. Ideally, marketing should result in a customer who is ready to buy. Adopting the marketing concept can often mean a total reorientation of the company so that its strategic planning starts by identifying opportunities created by the potential for more effectively satisfying customer wants, this then leads to the company's research and development capabilities being aligned to developing a range of services and a marketing mix which optimally matches the prior identified opportunities of the market. Drucker observed that marketing is so basic that it cannot be considered as a separate function, it is the whole business seen from the point of view of its final result, that is from the client's point of view.

11.4 THE MARKETING FUNCTION

There are two key decisions which are essential to marketing management: the selection of target markets which determine where the firm or practice will compete; and the design of the marketing mix (service, fee level, promotion and distribution method) which will determine its success in these markets.

11.4.1 Selection of target markets

In today's rapidly changing environment products, services and markets have a limited life expectancy, as discussed by Peter Lansley in the introduction to this Part. A firm which does not update and change its services in the markets, is unlikely to be successful for long. A major job of management is therefore to determine which markets offer the business opportunities for profit and/or growth in the future. Market research is the tool used to generate the information for reaching such decisions. Three areas for research are particular important. First, the firm will want to estimate the size and growth potential of alternative markets since in general it will prefer to operate in growth, rather than mature or declining markets. Second, it will wish to judge the strength and competition in candidate markets. How tough is the competition? Will it be possible to carve out a niche without strong reaction from existing firms? Thirdly, choice of target markets will be influenced by the fit of the market requirements with the firm's own strengths and weaknesses. A practice of quantity surveyors, for example, will probably wish to develop into markets which could use their skills in financial management and contract administration. In general, therefore, a company will seek opportunities in areas where it has some expertise, which will form the basis for a competitive edge.

After a broad market is identified comes the key task of market segmentation. Undifferentiated marketing, i.e. a single marketing mix offered to the entire market, is rarely successful because markets are not homogeneous but made up of different types of buyers with diverse wants regarding benefits, price, channels of distribution and service. A market segment is a group of potential clients with similar purchasing characteristics. In developing a marketing plan, the firm usually has to design appropriate offers for each segment if they wish to compete. Differential marketing is the policy of attacking the market by tailoring separate services and marketing programmes for each segment. Concentrated marketing is often the best strategy for the smaller firm. This entails selecting to compete in one segment and developing the most effective marketing mix for that sub-market. This could be particularly important in the construction professions, since the vast majority of firms and practices are small, with limited resources and rely on the skills and technical capabilities of specific people. A niche marketing strategy could therefore be particularly effective.

11.4.2 Developing the marketing mix

The marketing mix is the set of choices which defines the firms offer to its target market. McCarthy has popularized the 'four Ps' definition of the marketing mix – Product, Price, Promotion and Place. Potential clients in the target segment have a set of wants and by research and successful adaptation the firm will develop an offer to match them.

As emphasized earlier the marketing mix for one segment may need to be quite different from that of another; in tailoring its mix the firm will seek to offer one which target customers will see as superior to that offered by competition – this goal offering a marketing mix superior to competition is termed the Differential Advantage (DA) or Unique Selling Point (USP).

11.5 THREE KEY MARKETING PRINCIPLES

In analysing marketing problems and developing marketing plans three concepts play a central role: marketing segmentation, the differential advantage and positioning strategy.

11.5.1 Market segmentation

Pigou and Chamberlin developed the basic theory of market segmentation, showing that where different segments exist with separate demand functions, the monopolist would maximise profits by charging different prices to the two segments. There are many practical examples of this type of price discrimination, e.g. a public utility charging different rates for business and domestic consumers. The concept of segmentation in marketing however is much more general. The firm must recognize that clients differ not only in the fee they will pay but also in a wide range of benefits they expect from the particular service and how it is presented. The firm can discriminate not only in price but potentially in any of the four 'Ps'. For example, the market for a firm of general practice surveyors is made up of a number of segments including a segment for commercial clients, residential clients or industrial clients. The marketplace is further divided geographically since the commercial office market in Liverpool is quite different from that in London. Each segment is likely to have a different price elasticity but in addition each desires different approaches to providing the service. Similarly, the advertising and promotional strategies will differ in media and message between each segment. Segmentation is central to marketing because different client groups require different marketing mix strategies. The technique of segmenting a market also reveals profit opportunities and 'strategic windows' for new competitors to challenge established market leaders, as a market develops new segments open up and older ones decline. Most markets are a mix of fierce and relatively weak competitive segments, slow and high growth areas. The market-oriented firm will therefore be seeking to identify those dynamic segments offering the best growth and profit possibilities.

11.5.2 Differential advantage

Target market segmentation is, however, insufficient for strategic planning since in general other companies will also be carrying out the same analysis and be competing for any segment chosen. To be successful the firm must also develop a differential advantage (DA) or unique selling point (USP), which will distinguish their services from competitors in the segment. Only by creating such a differential advantage can the firm normally obtain high profits, since what they are effectively doing is creating a monopoly situation. In a market where no firm has a differential advantage then potential clients choose on the basis of price, and price or 'perfect' competition ensures that profits are pushed towards zero. The task of the modern business is therefore to seek what may be termed a 'quasi-monopoly' situation to make itself unique to clients so that they will not switch to competitors for minor price advantages. High profit firms are generally those which have succeeded in creating such preferences. These differential advantages may be obtained potentially via any element of the marketing mix creating a superior product to competition, more attractive designs, better service provision from the professional staff, more effective geographical capability, better advertising or selling and so on. The keys are understanding that 'advantage' is based upon research into what clients really value and that a differential is derived from an evaluation of competitive strategies and offers.

11.5.3 Positioning strategy

Positioning is the amalgam of these two earlier principles. Positioning strategy refers to the choice of target market segment which describes the potential clients that the business will seek to serve and the choice of differential advantage which defines how it will compete with rivals in the segments. Thus, Porsche is positioned in the prestige segment of the car market with a differential advantage based on technical performance and image. The appropriateness and effectiveness of the positioning strategy is the major determinant of a business growth and profit performance.

11.6 KEY ANALYSES FOR DEVELOPING A MARKETING STRATEGY

In formulating its choice of target market segment, marketing mix and differential advantage, the firm will focus its research in five major areas.

11.6.1 Market analysis

As described earlier, before committing itself further the company should assess the growth and profit opportunities likely to be open to it in the chosen market.

11.6.2 Customer analysis

The firm will need to research how the market is segmented, which of these segments are the most attractive, and what are the benefits desired by clients within each of these segments. Knowledge of such factors will be central to designing its service and offer. In addition, to develop its promotional strategy they will also need to determine who are the key people affecting product choice in the client organization and what type of buying process occurs before a decision is reached (e.g. how information is sought, how alternatives are compared).

11.6.3 Competitor analysis

Developing a differential advantage means making target clients an offer which is superior to any made by other firms in the market. Clearly therefore this strategy requires identifying who the competitors are now and whether new ones may emerge in the future. A firm needs to judge what their strategic objectives are, how their offer is perceived by potential clients, and how it may change in the future. Finally, it is crucial to estimate how competitors are likely to react to any strategic initiative on your part.

If you introduce new products or cut prices in an effort to gain market share, is this initiative likely to be nullified by speedy retaliation from competition?

11.6.4 Image

Because professional services are intangible, and production and consumption are simultaneous, a major point which needs to be addressed by the firm is that of whether the image of the firm corresponds with the quality of the service being provided. There are three types of 'image':

1. **Actual or current image**: what is the image that potential clients actually see in the firm?
2. **Wish image**: what is the image that your firm would like potential clients to see? Is it different from the actual or current image? If so, how are you going about changing it?
3. **Mirror image**: what image do you see yourself as being portrayed by your own firm? If it is different from the actual or current image, then corrective action must be taken, otherwise any promotional strategy is bound to be ineffective.

11.6.5 Economic and stakeholder analysis

In assessing a marketing strategy the firm will obviously wish to assess the financial implications and its impact on stakeholder groups of importance to the firm. For a business the marketing strategy will need to generate an adequate level of profit to satisfy the equity partners or shareholders and provide for the firm's continuing investment requirements. Since investment generally anticipates profit return, the cash requirement will also have to be considered. Can the business finance the marketing strategy out of cashflow, or will we need to borrow money on the promise of a future income stream? Other stakeholders or interest groups that may have an effect on marketing policy may include central and local government regulatory bodies and the relevant professional institutions, who regulate the conduct of their members through various laid down standards. The well-designed marketing strategy will have considered the impacts on all relevant parties.

11.7 MARKETING MIX DECISIONS

Market research and business appraisal are vital in order to establish what services the firm should be offering and to position itself against potential market segments, but to develop a differential advantage of its services is implemented by management decisions and the marketing mix. Important normative rules for optimising the marketing mix have been developed by economists. These rules stem from the model of the firm in which output and price are the only decision variables and where sales are optimized at the part at which marginal revenue equals marginal cost. Dorfman and Steiner extended this to include other marketing mix variables, namely advertising and product or service quality. The Dorfman–Steiner theorem showed that short run profits are maximized when the company balances lower prices, increased advertising expenditure and higher quality products in a relationship which includes elasticity of demand for their products or services. Where a product or service is effectively differentiated from its competitors, then the demand curve becomes increasingly inelastic, thus making the demand insensitive to changes in price, which is a position enjoyed by a monopoly provider.

Dorfman–Steiner theorem provides an important insight for designing the marketing mix and several studies have tried to estimate and apply this rule. In practice, however, its use is limited for a number of important reasons. In particular it is extremely difficult to estimate these marginal effects especially in oligopolistic markets where competitive reaction is a major dilemma and where many variables (advertising, intensity of distribution and price) interact with one another.

Also the firm is generally not seeking to maximize profits solely but may have a range of strategic goals (growth, risk avoidance, supporting complementary products in the firms range, etc.). Some of these issues become clearer when the various elements of the marketing mix are considered.

11.7.1 Product policy

Economic theory has little to say about product policy because the theory of consumer behaviour treats the products themselves rather than the benefits or characteristics they possess as a direct object of utility. Under such an assumption little can be said about the key questions of how one product competes with another or how one can develop a superior product. Recently, economists have tried to fill this gap by developing a new theory of consumer behaviour which defines a product as a bundle of characteristics which are the ultimate goal of the buyer. This may open the possibility of a much more fruitful set of insights into consumer behaviour.

Paradoxically, this notion that consumers are interested in the benefits provided by a product rather than the product itself has always been central to marketing theory. The 'new economics' mirrors an approach which has been long applied in marketing to the study of preferences. To a marketing manager a product is the constellation of benefits generated by the physical product, its design, features, packaging, style, and service support which, together, provide satisfaction to the consumer. It is often said that much of IBM's success was based on its recognition that their cost and complexity will make computers unattractive to industry executives. At the same time, IBM knew that these executives increasingly needed more effective systems for efficiently and rapidly handling and manipulating information. IBM's offer was not a computer but a management information system made up not only of the hardware but software, attractive input/output and related peripherals, technical support, training, installation, operational back-up and easy financial arrangements.

A product or particular service can therefore be seen as three concentric circles. These are the **core service**, the **actual service**, and the **augmented service**.

The core service

The inner circle is the 'core', i.e. the benefit that the service will be fulfilling for the client. No client employs a firm of architects as an end product in itself. The client generally wants a building to house a function related to his or her own business activities. Therefore the core product must be the design and possibly management of the building process.

The actual service

The firm of architects will be commissioned (by whatever decision process) and the relationship will start. This relationship, the meetings, the building design, the management of the process, and possibly the acceptance of the building on behalf of the client, will be the actual service. It should always be remembered that the client is rarely buying professional services in the construction and property industries, but they generally come with the product that he or she is actually buying.

The augmented service

Every firm of architects provides this same core and actual service. But in order to differentiate on what one firm provides for the client, it may be necessary to augment the service with other features or 'extras' such as a post-occupancy evaluation of the finished building, which will give the client a deeper or broader service. An example here is in the purchase of a motor car, where the core and actual products are reasonably obvious, but certain manufacturers augment the product by adding on insurance, free servicing for a period, or optional extras as standard.

For effective marketing it is also crucial to see how the benefits are perceived by consumers rather than how they are defined by production experts. Product positioning is a market research technique which seeks to elicit from buyers a description or map of how alternative services are perceived. Positioning studies have of range uses of in planning: they show the strength and weakness of the various brands along the dimensions which are important to consumers, they show how closely competitive brands are seen and they indicate where different segment preferences are. Such insights have obvious implications for repositioning the service by quality changes, design, packaging or advertising, modifications and also for new service development.

Positioning models suggest seven alternative strategies a firm can pursue by either modifying their product or by persuasive communications:

1. Developing new services.
2. Modifying existing services.
3. Altering beliefs about the firm's services.
4. Altering beliefs about the competitor's services.
5. Altering the importance attached to the individual characteristics.
6. Calling attention to neglected characteristics.
7. Shifting consumer preferences.

11.7.2 Pricing policy

The economic theory of price shows that profits are maximized when prices are set to equate marginal revenue and marginal costs. An obvious practical problem with this is that other variables usually affect the demand for a firm's services both independently and interactively with price. For example, some studies have shown that advertising increases price elasticity. In principle, modern econometrics provides methods of estimation in such circumstances and there have been a number of successful empirical applications published.

In general, however, few professional firms explicitly follow the economic model in developing a pricing policy. Most, quite naturally, find estimating the parameters of the demand function too difficult, time consuming and expensive. In addition, with such a rapidly changing professional and commercial environment few would expect such parameters to be stable in practice and therefore management of price tends to be on the basis of more intuitive judgements about the nature of cost, demand and competition. Pricing is also significantly affected by the firm's objectives. Where the firm has, for example, a long and strategic goal of winning a dominant market share then its price is likely to be significantly lower than for a firm aiming to maximize current profits.

In setting prices the professional firm has to consider not only its own profitability and the reaction of buyers but other parties too. Any other firms or people used in the company's final product, such as advertising agencies, printers and referrers, etc. must be given an adequate profit margin to ensure that they have sufficient incentive consistently to give quality service. The firm will also have to consider the likely reaction of competitors to a switch of pricing policy as not only are existing competitors affected by this pricing policy, but the level of profit margin is likely to influence the entire range of new competitors. From time to time, Government agencies also affect pricing policy for anti-inflation or anti-monopoly reasons.

Studies show that almost all professional firms establish their price on the basis of cost plus. Price is determined by adding some fixed percentage to total unit cost. Unfortunately such findings do not really tell us much without a theory of how the mark-up percentage is determined. A major determinant of the potential mark-up is the service's differential advantage, the greater the perceived value it has over competitive products the more the clients will be willing to pay. The most sophisticated marketing oriented companies like those in the computer industry calculate this perceived value to the consumer in considerable detail before making a pricing decision.

The McKenzie management consulting firm advocate calculating this value by comparing the product's total cost and benefit with those of a reference product, e.g. the market leader. For example, consider a company seeking to price a new service, for example post-occupancy evaluation mentioned above, in a market where the market leader is a competitive firm. Clients of the market leader pay say £3000 for the basic evaluation, another £2000 for certain extra services, such as a heating installation analysis, and over the next few years spend another £5000 in associated costs, making a total cost of £10 000. Suppose our new service has features which lower start-off and post-purchase costs to £4000, yielding a £3000 life-cycle saving. Then the economic value to the client of our service is £6000; that is, the client should be willing to pay up to £6000 to commission our firm. At £6000 there would be no incentive to switch from the market leader. Hence we may decide to price it at £4000 to produce a £2000 client saving incentive. This is all assuming of course that clients are price-sensitive, which is an issue we will return to in a later chapter.

When a company has designed a product and offer to match the wants of its target market segment, it then needs to communicate this offer to buyers and persuade them to try it. There are four main tools which we use to achieve these goals: advertising, personal selling, sales promotion and public relations. In general, before purchasing a product the buyer has to be brought through various stages of the communication process. First, he or she has to be made aware of the firm's existence and what it can do. Second, they should understand the benefits that the product offers. Third, they should be convinced that it will meet their wants and needs and, finally, they have to be brought to the point of making a positive purchase decision.

The different communications tools are frequently used to achieve specific communications goals. Advertising is particularly good at making the market aware of the firm and its services but it is usually far less effective than personal selling in closing the sale. Personal selling is usually a very costly means of creating awareness and comprehension, but more efficient at stimulating conviction and purchase. Sales promotion and publicity tend to be used far more widely for the marketing of professional services, mainly due to the regulations of the professional bodies regarding advertising. Additionally, because professional services are generally targeted at clearly defined market segments, techniques of mass advertising are generally ineffective, unless used in specific media, such as the professional journals or targeted newspapers or magazines, such as *Construction News, Property Week*, etc.

Here, we should note the key decision areas in promotion. In planning advertising the professional firm (often in conjunction with an advertising agency, if appropriate) will need to make five decisions. First, they will need a clear definition of the target market segment to which the promotion is to be directed. Next they will need to determine the most effective media (newspapers, journals, TV, radio, posters, etc.) for getting the message to the selected audience. Increasingly, agencies now use a computer program to seek an optimum combination of media, which will deliver the desired number of exposures to the target audience. The third area for decision is on the form and content of the message or copy to be expressed in the promotion. In general, an advertising agency will try to create a message which will express the differential advantage of the service in a manner which makes it believable, desirable and exclusive. The other two decision areas concern determining how much should be spent on promotion and how the investment should be subsequently evaluated and controlled.

Econometric studies have been widely used to estimate the pay-off of promotion and determine the optimum budget allocation.

11.7.3 Distribution policy

Distribution management is concerned with decisions on moving goods from the producer to the target consumers. Decisions about distribution channels are very important because they intimately affect all other marketing mix choices and because once made they are not easily changed. In principle, a manufacturer can choose between selling the goods directly to the consumer and using a variety of distributors and retailers.

The marketing distribution channel undertakes a number of tasks beside the physical transportation and storing of goods, and this is where the understanding of distribution is important for the professional firm, since no physical goods are produced, to be subsequently moved, stored and sold. Middle-men may undertake market research, promotion, pricing and sometimes negotiation with the client. Providers of services, such as insurance companies, often use professional firms as agents to perform some or all of these functions as this often leads to superior efficiency in marketing the goods to the customer. This, however, has implications for the professional firm in terms of business versus professional advice, as we shall see in a later chapter. Such intermediaries, through their experience of specialization and contacts, can often offer the insurance company more than they can achieve by going direct. In developing a distribution strategy the service provider will make choices about the types of intermediary to use, the number to use, specific tasks they are to undertake (e.g. advertising, pricing, acceptance) and the terms (profit margins, territorial rights, etc.) under which the intermediary will undertake these tasks. Central to the problem of channel management is the recognition that distributors are independent businesses with goals that are at least partially conflicting with those of the provider. Whether the provider can design the channel to meet their own goals depends upon the power they have and their ability to motivate the intermediaries to cooperate. Knowledge of these marketing issues has been enriched by a considerable number of research studies by behavioural scientists and economists studying market entry strategies.

In choosing distribution channels, the service provider will seek intermediaries which meet four criteria:

1. They should be oriented to serving the chosen target market. A firm requiring a market position in a particular geographical area, may choose to franchise its operations to an independent company rather than enter that market itself.
2. The firm may want distributors which help it to exploit its own differential advantage. If the firm's competitive edge is in sophisticated technology features offering cost savings to clients then they will

need partners capable of explaining these benefits to prospective clients.

3. Working with a particular channel must be economically rational for the firm. Direct selling with a company sales force is a powerful channel, but it is too expensive for companies without a significant market share.
4. The firm will naturally be influenced by the control and motivation of prospective intermediaries. A potential partner offering a wide range of successful competitive services might give insufficient attention to our firm's products.

Distribution channels once established are not easy to change, yet as the firm's circumstances or the market evolves, it often becomes necessary to adapt or even to revise radically the existing distribution channels. When starting out a firm may choose partners to market their services because this will reduce overhead costs, but if the company grows it is likely to become increasingly financially attractive to switch from using partners or franchisees to direct entry because the higher overheads can now be spread over a larger volume. In addition, channels like products and services are subject to life-cycles in that highly successful forms of distribution will give way to new forms more effectively geared to emerging markets and buyer behaviour. Variety stores, corner shops and small supermarkets have lost ground quite rapidly to superstores, catalogue showrooms and discount stores. Such forces mean that the firm must be continually monitoring the performance and prospects of its distribution arrangements and be prepared to adapt them when conditions change.

11.8 CONCLUSIONS

In today's rapidly changing and highly competitive international environment a business can be successful only if its offer matches the wants of buyers at least as effectively as its best competitors. Marketing management is the task of planning this match. It is based upon the analysis of customers, distributors and competitors, the selection of target market segments and the design of marketing mixes which will provide the firm with a differential advantage or unique selling point.

An organization's success in creating a differential advantage determines its competitiveness and profit performance in whichever market it chooses to operate.

The changing environment, changing wants, new competitors and technologies, different stakeholder pressures on the firm mean that a differential advantage is never secure. Change requires a firm to continually reposition itself by shifting from declining to emerging market segments and renewing its differential advantage by such measures as improving its service features, adapting new technologies or higher levels of service. Businesses which fail to develop such repositioning strategies gradually lose contact with buyers and give way to firms which are more successfully marketing oriented.

11.9 REFERENCES AND FURTHER READING

Bachner, J.P. and Khosla, K. (1981) *Marketing and Promotion for Design Professionals*, Van Nostrand Reinhold, Wokingham.

Baker, M. (1981) Services – salvation or servitude? *Quarterly Review of Marketing*, Spring, 8–18.

College of Estate Management (1989) *Marketing for Chartered Quantity Surveyors* (video), College of Estate Management.

Connor, R.A. and Davidson, J.P. (1985) *Marketing Your Consulting and Professional Services*, Wiley, New York.

Coxe, W. (1983) *Marketing Architectural and Engineering Services*, Van Nostrand Reinhold, Wokingham.

Dorfman, R. and Steiner, P.O. (1954) Optimal advertising and optimal quality. *American Economic Review*, December, pp. 826–36.

Fisher, N. (1986) *Marketing for the Construction Industry*, Longman.

Friedman, W. (1984) *Construction Marketing and Strategic Planning*, McGraw-Hill, London.

Golzen, G. (1984) *How Architects Get Work*, Architectural Press, London.

Jones, G. (1983) *How to Market Professional Design Services*, McGraw-Hill, London.

Katz, B. (1988) *How to Market Professional Services*, Gower, Aldershot.

Katz, B. (1988) *Selling Professional Services*, Gower, Aldershot.

Kotler, P. (1984) *Marketing Professional Services*, Prentice-Hall, Englewood Cliffs.

Kotler, P. (1994) *Marketing Management, Analysis, Planning and Control*, 8th edn, Prentice-Hall International, London.

Linton, I. (1985) *Promotion for the Professions*, Kogan Page, London.

Lovelock, C.H. (1991) *Services Marketing*, Prentice Hall, Englewood Cliffs.

McCarthy, E.J. (1978) *Basic Marketing: A Managerial Approach*, 6th edn, Holmewood, Illinois, USA, p. 39.

Monopolies Commission (1970) *A Report on the General Effect on the Public Interest of Certain Restrictive Practices in Relation to the Supply of Certain Professional Services*, HMSO, London.

Royal Institute of British Architects (1990) *Marketing*, R.I.B.A., London.

Teare, R. (1990) *Managing and Marketing Services in the 1990s*, Cassell, London.

Wilson, A. (1972) *The Marketing of Professional Services*, McGraw-Hill, London.

Wilson, A. (1984) *Practice Development for Professional Firms*, McGraw-Hill, London.

CHAPTER 12

Market analysis; the service/ market matrix

12.1 INTRODUCTION

In the wake of recent valuation-based litigation claims, the recommendations of the Mallinson Report and competition from other professions such as accountants and bankers, there is a clear need to review the service delivered by construction and property professionals. It is within this context that this chapter examines the scope for a more strategic approach as a way of improving the professional services by using valuation services of general practice surveyors as an example.

We will start by outlining the pressures which are leading to the adoption of a more strategic approach to the provision of professional services in general, and valuation services in particular. This leads into a discussion of the insights which can be provided by developing a more strategic perspective. This strategic perspective is based on understanding the market in which the service is provided and on devising strategies by which the service might be enhanced. The latter half of the paper explains the way in which some recent research attempted to apply this more strategic analysis to the provision of professional property services.

12.2 A STRATEGIC PERSPECTIVE IN THE PROVISION OF VALUATION SERVICES

12.2.1 Professional services and 'strategy'

Strategic activity can be defined to include those activities which relate to the identification and implementation of major issues that affect the whole of the organization and its long-term relationship with its business environment. As such, strategic issues include:

- the market segment that the organization should work in;
- the type of clients that the organization should be working for; and
- the range of services that the organization should offer.

In many areas of service provision, inadequate attention has been placed on the strategic context within which the service is provided. This lack of attention has arisen for two main reasons. First, the tendency of many service providers to concentrate on producing their services at the expense of seeking to understand the broader context in which they operate. Second, the perceived difficulties of identifying clients' requirements of services compared with requirements for physical goods. In particular, services tend to be less standardized, more intangible and more perishable than physical goods and their characteristics are perceived to generate difficulties in measuring the market and comparing service providers.

Over recent years, and during the 1980s in particular, there has been growing pressure to rectify this lack of attention paid to the strategic context in which professional services are provided. Fundamentally, this pressure has arisen from the increasingly competitive nature of professional services markets. This growing competition is, in turn, based on a series of pressures:

- Deregulation: the relaxation of regulations regarding promotion and business development in many professions.
- Client sophistication: leading to demands for more innovative and complex professional services.
- Rapid change in the business environment: changes associated with technology, productivity and internationalization are occurring with increasing rapidity.
- Oversupply of professionals: based on increased numbers, maturing markets and productivity improvements.
- Increased dissatisfaction with professionals.

Within the context of these pressures, professionals are becoming increasingly market-oriented and competitive in the service they provide. This competition has led to firms seeking opportunities to develop new markets and services to meet the requirements of their clients.

12.2.2 The strategic context and the surveying profession

The surveying profession has, over recent years, followed this trend and started to recognize the importance of the strategic context in which it operates. The pressures on the surveying profession have been similar to those facing other professions such as accountants, solicitors and management consultants. Two particular pressures for change relate to the increase in size of surveying practices, and growing competition in the provision of services. The past 15 years have seen considerable increases in the size of individual surveying organizations through processes of expansion, merger and take-over. Despite this growth, the approach to management and the nature of management skills in most surveying practices has not advanced greatly on those associated with the traditional partnership structure. Such traditional approaches to management have come under increasing criticism as a means of exploiting change in the market for surveying services.

The widespread redundancies associated with the recession, the likely oversupply of surveyors during much of the 1990s, and competition from other professions such as accountants and solicitors, has raised questions over the likely role of surveyors in coming years. In response, the profession as a whole has sought to increase competitive position in relation to other professions. As concluded in the Lay Report (1991):

> A service must be provided which is market demand led, rather than one arising from historical convention for the convenience of the supplier. We must produce a service, and the skill to provide it, which is relevant, which is within our area of competence, and one which is responsive to the needs of market demand.

A range of other research reports have identified the need to pay more attention to strategic issues in the provision of surveying services. The MAC report was particularly innovative in analysing the scope for surveying firms to generate competitive advantage in responding to the needs of the client. The CEM research addressed the issue in respect of training and developing the skills of property professionals in order for them to be able to compete more effectively in the changing environment of the 1990s.

Despite the undoubted progress within the profession, there remains a paucity of systematic studies, certainly in a published format, into the scope for developing a more strategic approach to the provision of different surveying services. In order to tailor a service to meet the needs and requirements of the market, it is important to have a clear knowledge of the needs and requirements of that market. Such knowledge can be gained by carrying out detailed research within the market, and it is within this context that the research presented in this paper was initiated.

12.2.3 Relationship with other work: the Mallinson Report

The bulk of existing valuation research is focused on the technical aspects of the activity. Over the past 2 to 3 years there has been growing pressure from within the surveying profession to understand valuations in their broader context. In particular, there has been growing concern to ensure that valuations accord more closely with the needs of the client. Various commentators have complained that the definitions of value contained in the Red Book are used as a shield by valuers and that they are unwilling to depart from these definitions for fear of later negligence litigation.

As stated by Jonas:

> To many clients, commissioning a valuation is no more than an expensive way of having someone else read the property press for them, with much of the real effort going into forming the exclusion clauses.

Such criticisms, and the rise in the scale of valuation-based litigation claims, led to the setting up of a RICS Presidential working party to examine the requirements of clients when they commission valuations. The findings of the working party, which was based on the views of over 100 clients, sought to assist the profession in attaining greater credibility and clarity in the provision of commercial valuations.

The focus of the report was in identifying areas for improvement in conducting valuations. Many of the recommendations, such as the use of discounted cashflow techniques, the wider availability of data and the inclusion of broader economic and price factors, related to the technical aspects of valuation. Beyond these recommendations, however, the report identified the central need to demystify the valuation process, largely by ensuring that both the client and the valuer have a full understanding of the process. Within this context, the present research provides a useful progression on the recommendations of the Mallinson Report by developing a strategic perspective to the provision of valuation services.

12.3 DEVELOPING A STRATEGIC APPROACH

12.3.1 Analysing the market for valuation services

The development of a strategic approach involves a range of management issues. It includes decisions about which services to provide in which markets and where to be located, as well as decisions about pay systems, production processes, organizational structure, management style and management promotion.

For our purposes, attention is placed on the strategic analysis of the market in which the service is provided in terms of the attractiveness of that market and the competitive position of different operators in the market. This analysis lays the basis for evaluating the options for developing services. Other issues, such as strategy implementation, are obviously of critical importance, but are not addressed here.

12.3.2 Understanding the competitive position

Perhaps the most important issue to be addressed in a strategic review of any market is the issue of 'competitive advantage'. In simple terms, competitive advantage comprises two components:

1. The attractiveness of the market in which the service is provided, in terms of market size, growth-rate, profitability, as well as social, legal and technological issues. A critical dimension of market attractiveness is the client or customer perceptions of the services provided and the willingness, or otherwise, to pay for the service.
2. The competitive position of firms operating in the market. This competitive position relates to issues such as competitive ability, market share and knowledge of customers and markets. An understanding of both components provides a basis for measuring the competitive position of specific markets or of individual companies. These two components are often portrayed as a matrix (Figure 12.1).

Within this matrix, the status of individual firms or services can be presented relative to their competition. Such a matrix can help in channelling resources to those businesses which contain medium-to-high industrial attractiveness with an average-to-strong competitive position, the belief being that these represent areas where there is a higher probability of an improvement in performance.

In Figure 12.1, the circle indicates that a hypothetical firm is competitively advantaged in an unfavourable market. The arrow indicates the direction in which the firm is forecast to move. Such a representation seeks can also provide insights into the relative performance of the business, and to provide a basis for comparing alternative

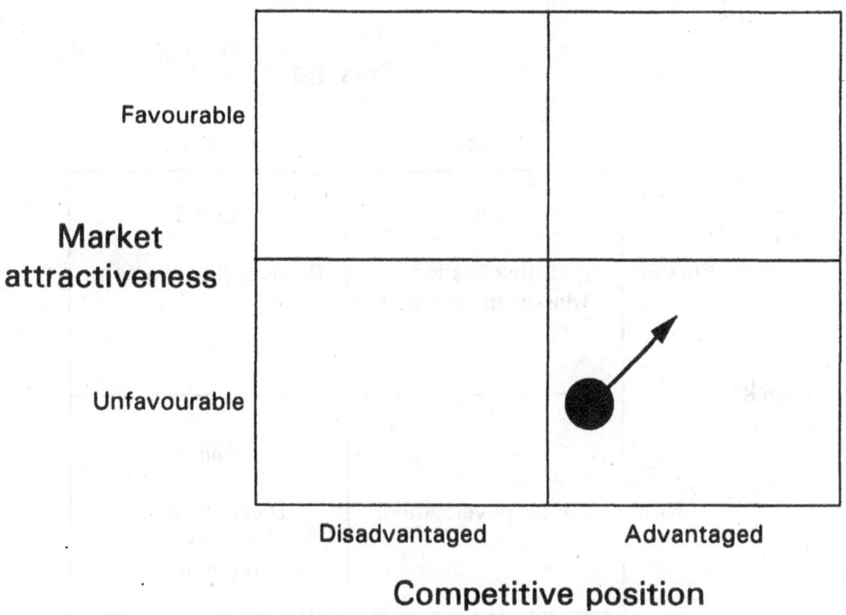

Fig. 12.1 Understanding the competitive position; market analysis.

strategies. As such, it can be used in the process for evaluating options for the development of the service.

12.3.3 Options for developing the valuation service

Anecdotal evidence suggests that valuation services might be developed in a number of ways ranging from fee cutting to specialisms in specific market sectors. In addition, it is possible for individual firms to increase the market for their services by providing the services to a new range of clients, either geographically or by activity. A useful approach to conceptualising this range of ways of developing valuation services is summarized in the 'service/market' matrix.

This matrix (Figure 12.2) shows the existing and potential new services along the horizontal axis, and the existing and potential market or clients down the vertical axis. The matrix is useful as it provides a framework for categorising the ways in which valuation services could be developed in order to generate a competitive advantage.

Within the matrix, Cell 1 is characterized by the firm's existing services which are provided to existing clients. Within this cell there is potential to **consolidate** the services provided in order to improve certain aspects such as quality or cost control. Consolidation might be associated with **market penetration** – in other words, an increase in market share at the expense of competitors. This market penetration might take the form of cross-selling of services to existing clients. For

Market \ Service	Present	New
Present	Cell 1 Consolidation Market penetration	Cell 3 Product development
New	Cell 2 Market development	Cell 4 Diversification related unrelated

Fig. 12.2 The Service/Market matrix.

instance, a property company client for whom the firm conducts asset valuations might also be interested in rating advice or in more strategic advice on the management of their assets.

Cells 2 and 3 in the matrix indicate development into new markets and new products respectively. In the case of Cell 2, **market development**, there is a development into new markets or identifying new clients for services the firm already provides. This might, for instance, involve moving into new market sectors or geographical areas. In all cases the rationale is one of further exploiting a company's current range of services by opening up new market opportunities.

In the case of Cell 3, **product development**, there is a development of new services to satisfy the requirements of existing clients. New services relate to enhancing the services provided by the organization. Innovations might involve minor modifications to existing services or the offering of new services.

The final sector of the matrix, Cell 4, **diversification**, indicates areas where the firm has neither a client or a service. Diversification into such areas is likely to involve a high-risk strategy, although this risk is lower where embarking upon a related diversification. Related diversifications might involve forward or backward integration to enable cost-savings, control of markets of the improvement of information.

The two approaches to market analysis provide useful frameworks within which to develop a strategic perspective on the provision of valuation services. Together, they could be used to evaluate a specific range of services provided by individual firms to a specific range of clients. This evaluation could lay the basis for enhancing both the services and the client base. The approaches could also be used on an industry-wide level, to demonstrate the overall attractiveness of the valuation market and the differentiation strategies pursued by different practices. In turn, the frameworks might help identify additional services or clients which might be exploited in order to increase business activity.

12.3.4 Bristol case study

Introduction

The scope to develop a strategic analysis of the valuation service market was explored for the case of the Bristol region. The selection of a specific geographic sub-market enabled a study of providers and consumers of the valuation service. The study was centred on the dimensions of market attractiveness and competitive advantage outlined above.

There are likely to be marked differences in the nature of valuation services in a provincial city such as Bristol compared with London. These differences, which are based on the relative importance of institutional and major corporate clients to London valuers, are likely to have implications for relationships between valuers and their clients. As such, the findings of the study are unlikely to reflect the experience in London. Against this, however, is the likelihood that the nature of relationships between valuers and their clients in Bristol is similar to other UK provincial centres and, as such, the conclusions of the present research should have implications for the rest of the UK.

Sample of surveying practices

The sample of surveying practices was based on the larger and medium-sized practices for four main reasons:

1. The larger firms are more likely to have specialized valuation services, as opposed to small firms where there will be little specialism with one or two staff members providing a range of services. As such, the staff in larger firms are more likely to be aware of the role of valuation as a discrete service within the firm.
2. Within the large to medium-sized firms, procedures and systems for carrying out functions are more likely to be standardized and thus offer scope for productivity improvement.
3. Larger firms are more likely to have a stated marketing strategy on how to improve business compared with small firms where informal networking is likely to be of greater importance in gaining new business.
4. Larger firms account for a large number of practising surveyors, with the largest 21 surveying companies identified in the 1993 EDB survey employing over 5000 professional staff.

There are difficulties in identifying medium-sized and large surveying practices at a local level due to the shortage of data on size and turnover on a geographical basis. The RICS and other sources such as EDB publish information on the size of individual organizations at a national level. The RICS also publishes a geographical directory which reveals that there are in the order of 50 surveying practices in the Bristol area.

The only information on the size of practice on a geographical basis relates to the number of partners/directors in specific organizations. This information on size is, however, misleading as certain entries for Bristol are based on the number of partners at the national level, while others identify the number of partners in the Bristol office. It was, necessary, therefore, to adapt a qualitative basis for selecting the firms and this involved discussions with valuers in Bristol and in-house knowledge within the School of Valuation and Estate Management. On this basis, 15 large and medium-sized commercial practices in the Bristol region were selected for the study.

Client sample

The valuer research was conducted before the client research. This revealed that the banking and corporate sectors together accounted for nearly half of all valuation business for the sample firms. Financial institutions were found to be a relatively small source of business, with such organizations tending to have national and international property portfolios which, in turn, require the services of national practices. Property companies also generated a relatively small amount of business, but the localized nature of their operations meant they were also included in the sample.

The relative importance of these groups and the localized nature of their operations were the reasons for concentrating on the bank, corporate and property company sectors in researching the views and requirements of clients. The importance of these clients means that their requirements need to be addressed by individual surveying practices and the surveying profession as a whole if opportunities for exploiting the market are to be taken. The focus on these groups also provided a sufficiently broad range of potential respondents to the questionnaire.

The selection of the sample was carried out on a semi-random basis which involved two stages. First, it was necessary to identify the number of organizations in the Bristol Region in each of the three groups namely, banking, corporate and property companies. Second, it was necessary to select a representative sample from each of these groups. The methodology is explained in more detail in an earlier publication.

In summary, a total of 104 questionnaires were sent out, 12 to the banking sector, 27 to property companies and 65 to the corporate sector. In overall terms, the response to the questionnaire was good, with a valid response rate of just over 24%. This was particularly good for an unsolicited questionnaire raising sensitive business issues such as costs and decision-making strategies.

The information provided by the two questionnaires was used as a basis for measuring the market for valuation services in the Bristol region. The main findings of the research are presented in the following section.

12.4 THE ATTRACTIVENESS OF THE VALUATION MARKET

The attractiveness of the market could be measured on a range of dimensions, including the value and profitability of the market, structural pressures facing the market and so on. The present research sought to capture the attractiveness of the market by focusing on four main dimensions:

1. The value of the market.
2. Reasons for commissioning valuations.
3. Client perceptions of valuers.
4. The scope for differentiation in the provision of valuation services.

12.4.1 The market for valuation services

The central measure of the value of the market for valuation services is the fees generated by the activity. The activity itself is an essential skill of the professional valuer and estates manager. Property valuations are explicitly or implicitly essential in many aspects of property advice. This applies whether the advice relates specifically to the giving of valuations or more indirectly in providing agency or development advice.

The pervasive significance of valuations in providing professional property advice is the major reason for its relative importance to surveying practices in terms of fee turnover. It is difficult to arrive at precise estimates of fee generation from valuation services due to the paucity of information on many aspects of professional property practices. The absence of much of these data complicates the analysis of the profession in general and the provision of valuation services in particular.

The most authoritative estimates of the scale of the market for valuation services have been provided by the MAC and the Avis and Gibson research of the 1980s and the annual Economic Development Briefing Chartered Surveyors Survey. The MAC research estimated that around 10% of fee income for services to existing property related to valuation services. This is likely to underestimate total fee income from valuation services as it was based only on institutional and property company holdings. As such, it excluded fees generated by corporate users of property, banks and the public sector, as well as fees associated with property development. The EDB survey estimates that total fee income from Britain's top 3000 property clients amounted to just over £840 million in 1993. Of this sum, it was estimated that about 7% (nearly £60 million) was generated for valuation services. Once again, this is likely to be an underestimate as it excludes expenditure by the public sector on surveying services.

The Avis and Gibson research adopted a different approach. Rather than estimating the amount of fees generated by valuation services, they identified the four most important fee-earning activities for 200 general practice firms. Valuations were clearly the most important service, with over 75% of firms ranking valuation as one of the four most important fee-generating service.

Problems in estimating the value of the market for valuation services are apparent at the local, as well as the national, level. For this reason, the scale of the market for valuation services in Bristol was taken to be indicated in two ways. First, the fees generated by such work and second, expenditure on valuation services by different groups of client.

On average, valuation services were found to generate just over 20% of the fee income of surveying firms within the Bristol region. This figure is higher than the national average of 7–10% provided by the research discussed above. This difference might be explained by three factors. First, previous research is likely to represent an underestimate of total fee income. Second, the sampling method of the present research which selected firms on the basis of their size and their perceived importance as providers of valuation services. Third, and as supported by the Avis and Gibson research, a lower proportion of income can be expected from agency and consultancy fees in a provincial market such as Bristol compared with the London region.

Fees generated from valuation services appeared to have fallen sharply during the recession, with the fee charged for a typical valuation falling by nearly 40% between 1990 and 1993. This demonstrates that valuers had responded to changing market conditions and supported widespread anecdotal evidence.

On average, the valuers felt fees would recover by about 20% over the 2 years to the end of 1995. This would have left fees considerably below the levels of 1990. However, this average movement in fees reflected mixed perceptions over the outlook for the following 2 years. Nearly two-thirds of respondents felt fees would increase by between 12 and 40% over the 2 years, in contrast to the remainder who felt fees would continue to fall or remain at 1993 levels. This uncertainty over future fee levels was reflected in the written comments received from respondents which also indicated the widespread hope that there would be an upturn in the market with a consequent increase in the amount of business.

In terms of expenditure on valuation services, the research found that the bulk of clients generated relatively low levels of fee income for valuers, with over half the respondents annually spending below £5000 on valuation services (Figure 12.3). Of these low spenders, two-thirds were in the corporate sector – which might be expected as such organizations would tend to have valuations carried out on a fairly infrequent basis. The other low spenders were property companies, reflecting the small size and scale of the property companies included within the sample. In contrast, less than 15% spent more than £20 000 per annum on valuation services. Each of these high spenders were banks.

12.4.2 Reasons for commissioning valuations

Beyond its value, an important dimension of the attractiveness of the market is the reason for which valuations are commissioned.

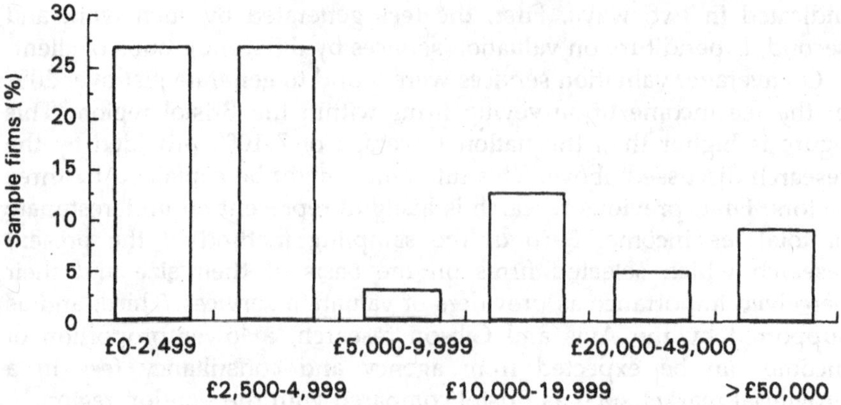

Fig. 12.3 Annual expenditure on valuation services, percentage of sample firms.

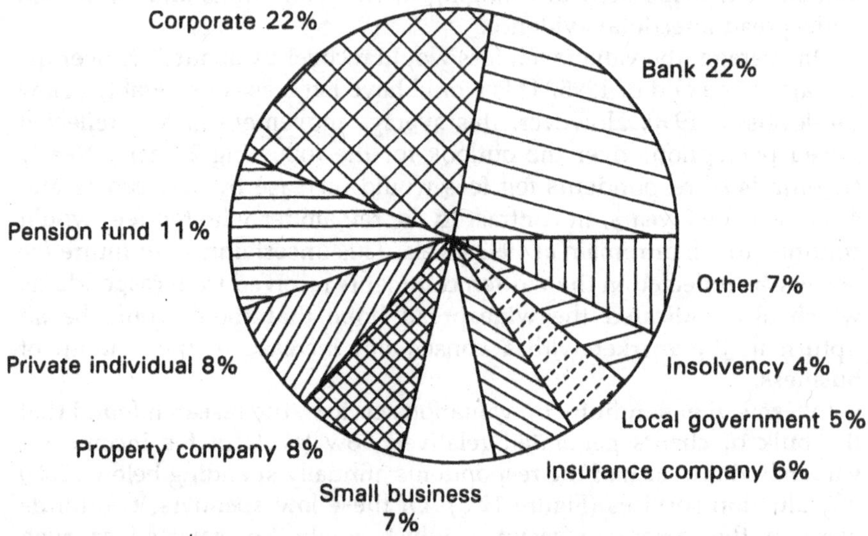

Fig. 12.4 Valuation business generated by different types of clients.

An understanding of the reasons is critical in anticipating future changes in requirements for the service.

Before examining specific client requirements, it is useful to explain the main sources of valuation business. In terms of fee income, the corporate and banking sectors together accounted for nearly half (44%) of all business for the sample of firms (Figure 12.4). Pension funds and insurance companies accounted for a further 17% of total business, with the remaining business being generated by a range of other clients.

These results reflect the localized nature of the market which the Bristol firms serve and the relatively small amount of 'prime' property in the region. Both these factors explain the relatively low level of institutional clients, with such institutions being more likely to have nationally managed and valued portfolios.

The actual purpose for which valuations are required varies according to the nature of the client and their specific reasons for commissioning valuations (Figure 12.5). For example, property companies have a statutory duty to have a full independent valuation of their portfolios carried out every 5 years, while non-property companies are not bound to have their property revalued at any regular intervals, although many companies have valuations carried out for their own purposes.

The different pressures behind the commissioning of valuations by various groups of client is a critical factor determining the attractiveness of the market. An understanding of these pressures and requirements enables assessments to be made of likely future growth and decline in fee generation and in the nature of client requirements.

12.4.3 Client perceptions of valuers

A related issue to the reasons for commissioning valuations is the way in which clients perceive valuers as providers of the 'valuation service'. In recent years many commentators have suggested that valuers are losing business to other professions, and to accountants and lawyers in particular, due to the poor image of the valuation profession. An important issue, therefore, is the views of clients towards the service provided by valuers in comparison with other professionals. This is particularly relevant given that such professions have begun to pose a threat to the valuation profession as they offer themselves as being capable of providing strategic valuation advice as opposed to the more technical skills of the valuer.

In overall terms, the research demonstrated that clients have a positive perception of valuers on a range of aspects of service provision. Most fundamentally, all of the respondents stated that the service offered by valuation firms accorded with their requirements. When asked to compare the quality of service provided by valuers with that provided by other professionals, nearly two-thirds of respondents felt that valuers compared either 'well' or 'very well' with other professionals. The remaining third of respondents felt that valuers could be ranked as 'fair' while none of the respondents chose either 'poor' or 'very poor' as a response.

It has also been claimed that valuers are not offering value for money and that this is damaging their image. However, an overwhelming 97% of respondents felt they were getting value for the money spent on valuation services. The responses to these two last questions would

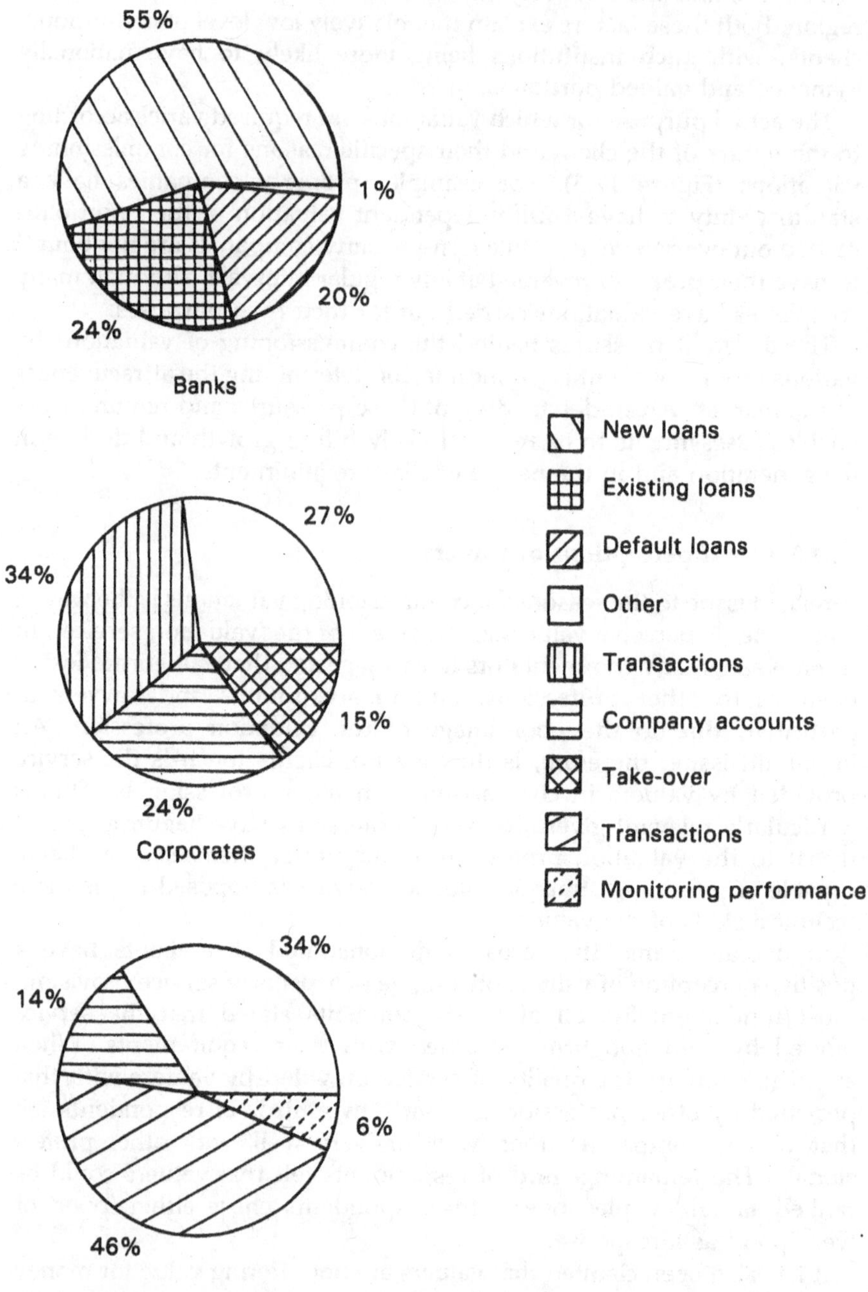

Fig. 12.5 Reasons for commissioning valuation services.

seem to refute allegations that clients of valuation services are deeply unhappy with the level of service provided.

Despite this positive perception by the clients, it is clear that the requirements of the clients are highly dynamic. Only one-fifth of clients had used only one firm of valuer in the previous 3 years. A surprisingly high proportion (36%) have used over five valuers over the same period. Clients appear to prefer to appoint valuers according to the nature of the valuation required rather than relying on a single firm of valuers to accommodate their requirements. This apparently high turnover in the use of valuers has implications for client retention strategies and demonstrate the need to understand the precise requirements of such firms.

12..4.4 Scope for differentiation

The final dimension of market attractiveness that was addressed in the research was the scope for service differentiation. There would appear to be a central tension in the provision of valuation services, between the need to comply with regulatory requirements and the ability to differentiate the service. On the one hand, there are considerable legal and regulatory constraints upon valuers in making valuations, most particularly in the form of the Red Book. On the other hand, there would appear to be scope to differentiate valuation services due to the range of reasons for conducting property valuations and the ability to vary the technical and professional approach to the valuation process.

Regulations influence the practice of making valuations and the basis upon which the valuations are made in a number of ways. Perhaps the most fundamental of these regulations is the Statement of Asset Valuations Standards first published by the Asset Valuations Standards Committee in 1974 and revised in 1990. These standards, which were established in order to provide a uniform standard for asset valuations, are:

> . . . primarily concerned with the bases of valuation to be adopted and the format of reports rather than laying down rules for the determination of values.

The Statements of Asset Valuation Practice provides guidelines for those valuations which may be included in published documents. They do not apply to other valuations which are undertaken for the private purposes of the client. Where valuations covered by the regulations are undertaken by professional property practitioners, it is necessary to observe and comply with the practice statements and guidance notes. Some of the guidance, such as that relating to the taking of instructions and the form in which valuations are to be presented to the client, is highly detailed. For instance, the appendix to the SAVP provides a checklist for the areas which should be considered by the valuer in making a valuation.

It is clear that legislation and regulations such as the SAVP generate a strong pressure for uniformity in the nature of the valuation service provided by property professionals. Despite this pressure for uniformity, there is scope for differentiating valuation services in two respects. First, in a technical sense in terms of the approach to determining values and, second, in the business sense in the way in which the valuation service is organized and provided for the client. These issues are explored further in the following section.

12.5 COMPETITIVE POSITION OF PROVIDERS OF VALUATION SERVICES

12.5.1 Introduction

A comprehensive analysis of the competitive position of service providers would involve a range of issues such as market shares, technological capabilities, the scope for service substitutes and so on. The present research did not seek to provide such a comprehensive analysis, but focused on two key issues relating to the rivalry among existing firms. The first of these is differentiation in service provision and the second is the organization of the service.

12.5.2 Differentiation in service provision

One of the main routes to superior relative performance is by differentiating the service provided in ways that are valued by the client. Such differentiation could enable the charging of a premium price by the service provider. The previous section explained that despite the pressure for uniformity, there might be scope for differentiation in the provision of valuation services. This issue was explored by examining two distinct types of differentiation. First, in terms of the perceived specialisms and differentiation provided by the firms, and second, by the reality of the specializations in terms of techniques use and time spent on different types of valuation.

All but two of the firms perceived themselves as specialising in particular areas, and there appeared to be a wide range of specialisms, with no more than three firms specializing in the same area (Figure 12.6). This might suggest that the market was fairly sophisticated in that the providers of the services were, inadvertently perhaps, aware of what their competitors provided.

In a similar way, firms appeared to be 'well known' by clients for different reasons. In total, there were 14 different specialisms identified

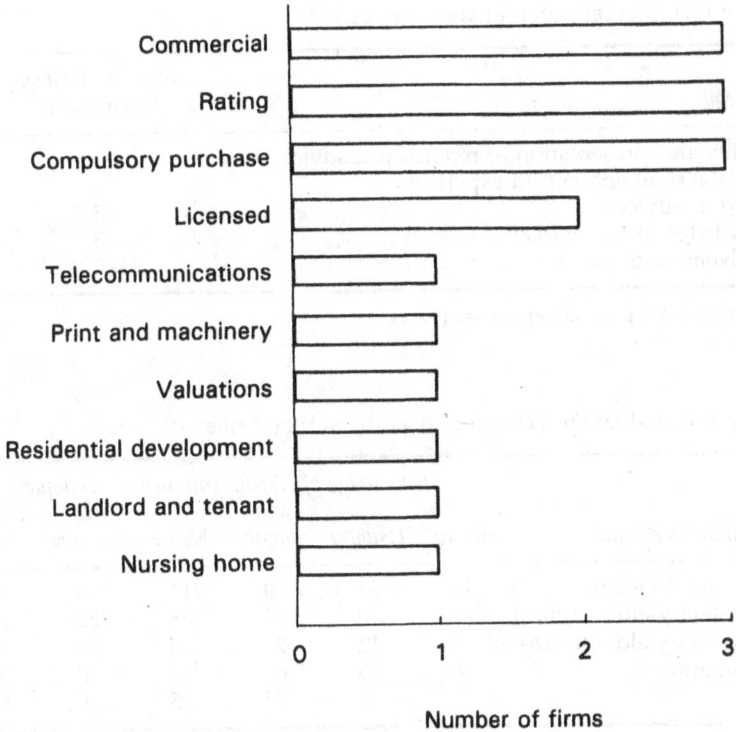

Fig. 12.6 Specialisms provided by surveying firms.

by valuers for which the firms were well known. For nearly three-quarters of these specialisms, only one firm identified themselves as being well known. For two of the remaining specialisms two firms were well known, and only for three specialisms were three firms well known. It was clear, therefore, that valuers believed there to be scope to develop specialisms, and this was confirmed by all but one of the respondents feeling it was possible to differentiate valuation services.

The reality of differentiation appeared to differ from these perceptions on a number of grounds. First, despite the claim that valuation services could be differentiated, firms sought to do so in a narrow range of ways. Specifically, the firms which felt it was possible to differentiate valuation services only identified five different strategies out of a total of 18. The most important of these were the quality and presentation of reports and advice, and the availability of specialized expertise (Table 12.1).

In one other respect – the valuation technique used in producing the valuation – there was also relatively low differentiation in practice. There was little differentiation between the firms in terms of the valuation techniques they employed, with the term and reversion

Table 12.1 Market differentiation strategies*

Strategy	*Number of times mentioned*
Quality and presentation of reports and advice	5
Availability of specialized expertise	4
Speed of advice	3
Knowledge of the market	3
Involvement of partners	2

*From Bristol region valuer survey (1994)

Table 12.2 Valuation techniques used by survey firms*

Valuation technique	*Percentage of respondents using technique*					
	Always	*Usually*	*Occasionally*	*Never*	*No response*	*Total*
Term and revision	25	63	0	12	0	100
Equivalent yield – vertical	12	0	25	38	25	100
Equivalent yield – horizontal	0	12	25	38	25	100
Layer/hardcore	0	25	63	12	0	100
DCF	0	0	75	25	0	100

*From Bristol region valuer survey (1994)

method being used 'always' or 'usually' by all but one of the firms (Table 12.2). This pattern, which was similar to the one revealed by Crosby in 1991, revealed the low technical sophistication in the provincial market which, in turn, probably related to the different client base in a city such as Bristol compared with London.

These results demonstrate that although the firms perceived themselves to differentiate their services, the actual strategies adopted were fairly narrow. This might have indicated the limited scope to differentiate valuation services or, on the other hand, it might have indicated that this scope was not being exploited. It is also worth noting that the strategies used to differentiate their services were not particularly dramatic and appeared to indicate a lack of innovation on the part of the firms.

12.5.3 Producing the valuation service

An alternative route to maintaining competitive advantage relates to the cost of the service provided:

> Low cost can enable the firm to compete on price. . . . It can also generate profits that can be reinvested to improve the product quality whilst charging the same price as the average in the industry. So it is not low cost as such that confers competitive advantage, it is the consequences of low cost that improve competitiveness. (Bowman, 1990)

The cost of providing the service depends, to an increasing extent, on achieving high productivity. For this reason, the research explored the ways in which the service was produced and strategies that were being adopted to improve the efficiency of service provision. The production of the service relates to all stages including the generation of new business, the receipt of instructions, the conducting of the valuations and the client's payment of the fee for the completed service.

There was general consistency between the firms in the time spent on different aspects of the valuation process. Over half of the total time involved in providing valuation services related to the three activities of conducting property inspections, identifying details of comparable rents, and writing valuation reports (Figure 12.7).

This breakdown (Figure 12.7) is of great interest for two main reasons. First, it clearly demonstrates the range of practical activities which need to be carried out in producing a valuation report. This concern to cover the practical aspects of the valuation will tend to drive the valuation process and it might explain the concern of practising valuers over the

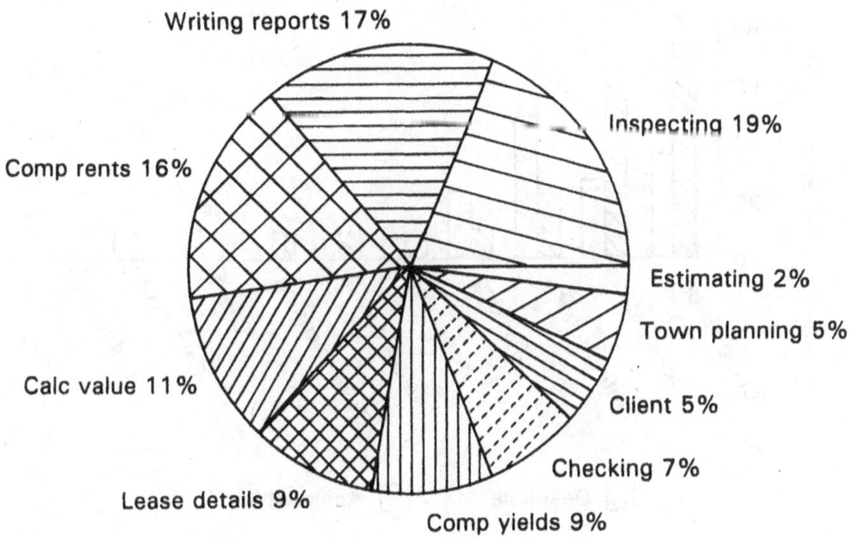

Fig. 12.7 Time devoted to various aspects of the valuation process.

attention paid by academic research on the relatively narrow area of valuation techniques.

More importantly, the focus on practical activities appears to have been deflecting the attention of valuers away from the needs of the client. There was a surprisingly low amount of time spent on liaising with clients, with, on average, less than 5% of time involved in this activity. This was worrying given the importance of ensuring that client requirements are fully understood.

All but one of the firms claimed they had sought to improve productivity over the previous 3 years. There was also considerable consistency in the strategies which the firms had pursued, with none of the firms seeking to raise productivity through reducing professional or non-professional staff, or through reducing profits. In contrast, productivity gains had been introduced by increasing the workload and standardising the service

When questioned on those parts of the valuation process where efficiency gains would be desirable and/or achievable, the majority of firms, not surprisingly, expressed a desire to improve efficiency in those areas which account for the largest proportion of overall time (Figure 12.8). In particular, the areas of comparable rents and yields which, together, accounted for over one-quarter of the total time in conducting valuations were seen to be areas where efficiency improvements were desired. It was also these areas where there was greatest discrepancy between the desire to achieve efficiency improvements and the ability to achieve them (Figure 12.8).

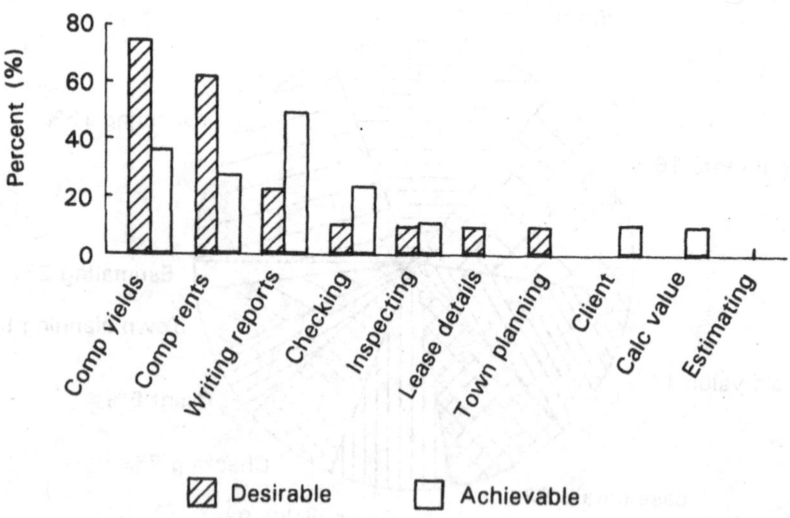

Fig. 12.8 Improving the efficiency of valuation services; desirable and achievable aspects.

This might indicate that it was these areas where greatest productivity gains could have been expected over the coming years, or, on the other hand, it might reflect the complexity of these areas which makes it difficult to standardize the process. There were signs that the introduction of standardized procedures could improve efficiency, although the small sample size means it is necessary to qualify these conclusions. Three firms stated that standardization was 'very important' in improving productivity, and a key area for this standardization would appear to be the collection of comparable rents. These three firms spent an average of 10% of valuation time on comparable rents, compared with nearly 20% for the remaining firms. None of the remaining firms felt that standardization was very important.

One of the main ways in which the efficiency of producing valuation service can be improved relates to the use of information technology. This was of particular significance in the Bristol area where there was a relatively low usage of information technology in producing valuations (Table 12.3). Of most surprise was that only one firm 'usually' uses valuation software to conduct valuations. The low use of such software might relate to the standard valuation techniques which were used by most of the firms (see Table 12.2). There was also a low use of FOCUS and this might be explained by the localized nature of the firms and the resulting detailed knowledge of the local market which the firms possessed. The one area in which there was widespread usage of information technology was in the use of in-house databases which all the firms 'always' or 'usually' used.

Table 12.3 Application of information technology in conducting valuations*

	Percentage of respondents using technique					
Type of technology	*Always*	*Usually*	*Occa-sionally*	*Never*	*No response*	*Total*
Valuation software	0	14	86	0	0	100
FOCUS	0	29	43	29	0	100
IPD	0	0	57	29	14	100
Property forecasts	0	0	42	29	29	100
In-house database	57	43	0	0	0	100

*From Bristol region valuer survey (1994)

12.6 OPTIONS FOR DEVELOPING THE SERVICE: SOME PRELIMINARY CONCLUSIONS

This chapter has provided a preliminary assessment of the scope for developing a more strategic approach to the provision of valuation services. The paper only covers a few of the key issues facing the valuation market. It should be taken to be indicative of the type of research that could be conducted on a firm-by-firm basis or at an industry-wide level.

It is only in recent years that valuers have come explicitly to recognize the importance of developing a more strategic approach. Pressures of competition and growing client sophistication have increased the importance of understanding client requirements, and the property profession is starting to respond to these pressures. To date, however, academic research and debate has tended to concentrate on the techniques used in valuations rather than the overall provision of valuations as a service to a client body. The present research has sought, in part, to rectify this gap in understanding by providing a more strategic review of the market for valuation services.

The research reveals a range of ways on which the attractiveness of the market might be measured. In terms of its value, the market is relatively unattractive, with pressures on fee-income and a large number of small fee generating clients. In addition, the market is highly dynamic, as indicated by the frequency with which clients change their valuer. Against this are more favourable features such as the range of different client requirements, in terms of the nature of clients commissioning valuations and the range of reasons for which valuations are commissioned.

In the face of the marked variations in client requirements, there appeared to be some differentiation in the valuation services provided. On the one hand, there was a range of firms, both large and small, providing valuation services. On the other hand, firms attempted to differentiate their product by developing expertise in specific areas with no more than three firms specialising in any particular area. Indeed, there was a unanimous belief among valuers that they, as a group, provided a range of specialist services.

The reality, however, appeared to differ from these perceptions, with limited differentiation in practice. Specifically, there was general consistency in approach to key issues such as the checking of valuations, the use of information technology and the desire to increase productivity.

In overall terms, clients appear to be generally satisfied with the service provided by valuers, but there is clearly a need for, and scope to, enhance the quality of service provision. The central dimension to such enhancements is a desire to ensure that valuation services are tailored to the requirements of various groups of client. Such enhancements should, in turn, enable individual firms to increase their market share, whether by more effectively retaining clients or by encouraging potential clients to use their services. They should also enlarge the scale of the market as a whole by increasing demand for valuation services beyond traditional reasons for commissioning valuation into the more proactive aspects of property management.

The research findings demonstrate the complexity of the market for valuation services. It is clear that the provision of this service requires the development of considerable technical and professional skills. The emphasis on these skills appears to have reduced the ability of firms to identify and exploit changes in the market for their services. Within this context, it is possible to identify a series of strategies which might be employed for developing valuation services, and these strategies are summarized in Table 12.4. The table demonstrates that firms were engaged in strategies to develop valuation services, albeit on an informal basis. It also shows the areas in which there might be further scope for differentiating services.

The need to consider such strategies is of growing importance due to a range of factors such as the continuing pressures on costs and the

Table 12.4 Strategies for developing valuation service in the case of Bristol

Strategy and key component	*The state of the market for valuation services in the Bristol region*
Market penetration (exploiting potential leads)	Low level of market research to identify client requirements Low level of cross-selling Most market penetration conducted on an informal basis
Consolidation (cost control)	Rather than redundancies and reducing the quality of staff, firms have increased standardization in conducting valuations Evidence that standardization can improve productivity in key areas Considerable flexibility in providing valuation service, with costs per typical valuation varying during the property cycle
Market development (acquiring new clients)	Two-thirds of clients are existing clients Dependent on a narrow range of clients (corporates and banks) although there is a potentially broad range of clients within the Bristol region Low level of research to investigate the requirements of existing or potential new clients
Service development (developing new services)	Limited application of information technology Similar valuation techniques employed by all firms Valuation techniques perceived to be relatively unimportant to clients by valuers

national debate over the nature of the valuation service required by the client. In addition, the Bristol firms faced growing competition from London-based, national firms. The client base of the Bristol firms was highly localized and one of the great strengths of local firms appeared to be their excellent local market knowledge. A number of respondents felt this competitive advantage over their national competitors was under threat. In particular, there was concern that if the market continued to improve, larger national firms might take the opportunity to set up provincial offices and thus take away the local firms' advantage in this area.

The threat to such activities is particularly acute given the dependence of local firms on a narrow range of client types and the different approaches towards valuation adopted by national firms. This threat is highly significant given the relative importance of the valuation service to local firms in terms of fee generation and the number of staff engaged in the activity. This chapter has sought to demonstrate that the nature of these threats and the development of strategies by which the threats might be tackled can be better understood by adopting a strategic approach to the provision of valuation services in particular, although the methodology and structure of the analysis can just as easily be used with other professional services within the construction and property industries.

12.7 REFERENCES AND FURTHER READING

Askham, P. (1993) Asset Valuation, in *The Best of Mainly for Students* (eds P. Askham and L. Blake), Estates Gazette, London, pp. 307–18.

Avis, M. and Gibson, V. (1987) *The Management of General Practice Surveying Firms*, Department of Land Management and Development, University of Reading.

AVSC (1990) *Statements of Asset Valuation Practice and Guidance Notes*, 3rd edn, Asset Valuation Standards Committee, London.

Baker, M.J. (1992) *Marketing Strategy and Management*, 2nd edn, Macmillan, Basingstoke.

Barrett, P. (1993) *Profitable Practice Management for the Construction Professional*, E. & F.N. Spon Ltd, London.

Bowman, C. (1990) *The Essence of Strategic Management*, Prentice-Hall, London.

Bowman, C. and Asch, D. (1987) *Strategic Management*, Macmillan, London.

Brett, M. (1993) What are valuations for? *Estates Gazette*, 30 October, Issue 9343.

CEM (1992) *The Skills Mismatch*, College of Estate Management, Reading.

Crawford, P. (1993) Valuation: what do clients want? *Estates Gazette*, August 28, Issue 9334, p. 20.

Crosby, N. (1991) The practice of property investment appraisal: reversionary freeholds in the UK. *Journal of Property Valuation and Investment*, 9(2), 109–22.

EDB (1993) *The Chartered Surveyors Survey*, Economic Development Briefing, Cinderford.

EDB (1994) *The Chartered Surveyors Survey*, Economic Development Briefing, Cinderford.

Forsyth, P. (1992) *Marketing Professional Services – a handbook*, Pitman Publishing, Financial Times Publishing, London.

Hicks, G. (1994) Valuations for Loan Security Purposes, Unpublished Address delivered in the Faculty of the Built Environment, University of the West of England, 26 February, UWE, Bristol.

Hobbs, P. and O'Leary, G. (1994) Client requirements from valuation services. Working Paper No. 38, Faculty of the Built Environment, UWE, Bristol.

Jonas, C. (1992) *Presidential Address*, RICS, London.

Jonas, C. (1993) Valuation: the way forward? *Estates Gazette*, 28 November, Issue 9347, p. 58.

Kotler, P. and Bloom, P. (1984) *Marketing Professional Services*, Prentice-Hall, New Jersey.

Lay, R. (1991) *Market requirements of the profession*. Report to the RICS by the Lay Committee, RICS, London.

Lewis, C. H. (1993) *Business Needs Us*. Address by the President, 6 July, RICS, London.

Lewis, C. H. (1994) Paper delivered at 'Valuation and Forecasting' Joint CULS and SPR Conference, 8 February, London.

MAC (1985) *Competition and the Chartered Surveyor: changing client demand for the services of the Chartered Surveyor*, Management Analysis Centre, London.

Mallinson, M. (1994) *The Report of the President's Working Party on Commercial Property Valuations (The Mallinson Report)*, RICS, London.

Morgan, D. (1993) Fee cutting, *Chartered Surveyor Weekly*, 7 October, pp. 28–9.

Morgan, N. (1991) *Professional Services Marketing*, Butterworth-Heinemann Ltd., Oxford.

Porter, M. (1980) *Competitive Strategy*, Free Press, London.

RICS (1994) *Geographical Directory 1994*, Macmillan, London.

Watkins, J., Drury, L. and Preddy, D. (1992) *From Evolution to Revolution: the pressures on professional life in the 1990s*, University of Bristol and Clerical Medical Investment Group, Bristol.

Whitmore, J. (1994) Mallinson Report bowls over industry. *Property Week*, 21 April, p. 4.

CHAPTER 13

The changing public sector marketplace and its implications for private sector firms

13.1 INTRODUCTION

Major changes in the public sector have meant that government bodies and local authorities are implementing large-scale and widespread tendering of property consultancy for contracts which will take effect in 1996. During the 3 years to March 1995, government departments have already exposed estates and facilities management services to the market with an estimated value of over £423 million. The total value of all construction related services within local government has been estimated to be in the region of £1.5 bn, which therefore represents some 28% of total workload and is consequently highly significant.

A considerable challenge is therefore being presented to private sector firms engaged in property consultancy. Tendering is also becoming more common in the private sector, but traditionally has not been the way that surveying firms, particularly those in general practice, have secured fee instructions.

Independent research conducted by the College of Estate Management during 1995 set out to examine the impact of public sector tendering on the market for property consultancy. The overall aim of the project was to recommend 'best-practice' approaches to developing beneficial client/consultant relationships within the competitive tendering framework. The study has concentrated on the tendering of general practice property work. However, many contracts let by the public sector are multi-disciplinary and involve several property specialisms, such as building design, construction, maintenance and management. The way in which the public sector packages and lets contracts for property advice could therefore have a significant impact on the private sector market and on the way that firms are managed and structured.

The main research report (published by the College of Estate Management in 1995) examines in more detail the wider issues relating to the way that tendering and contract arrangements are evolving in the context of a changing national economy. The findings discussed here represent a preliminary analysis of the research data and concentrate on issues related to managing the business unit, and particularly implications for private sector firms.

13.2 THE RESEARCH APPROACH

The research draws on the experiences and perceptions of public and private sector property professionals and managers who have had involvement with tenders for public sector property work. The majority had experience of fixed term contracts, usually of 3–5 years' duration and ranging in value from under £50 000 to around £2 million per annum. Structured face-to-face interviews were undertaken with 49 individuals as shown in Table 13.1. The first 12 interviews were also used as a pilot study to help design an attitudinal questionnaire completed by subsequent interviewees and three individuals who could not be seen in the time available.

A qualitative approach was chosen as the best method to gain cooperation from key individuals in relation to this commercially sensitive topic. Assurance of confidentiality was no doubt a factor in achieving a response to requests for interviews at a rate in excess of 90%

Table 13.1 Interviews conducted relating to procurement of property advice

Interviewee	*Private sector*	*Central government*	*Local government*	*All*
General practice surveyors	21	3	5	29
Other property surveyors	9	1	3	13
Other professionals	2	4	1	7
Total individuals seen	32*	8	9	49
Total number of interviews				43

*Included 23 firms of chartered surveyors ranging in size from two partners to 200 qualified property staff

13.3 MAIN ISSUES FROM THE RESEARCH

This section presents the main issues from the research within the context of organizational and legislative changes affecting the public sector. The choices and challenges for private sector firms are also examined in the light of other recent related research.

13.3.1 The public sector context

Interviews for the research deliberately involved people who had experience of contracts which had been tendered by both central and local government. There are two particular reasons for this which emerged from the pilot interviews. Firstly, property consultants see more similarities than differences in the tendering procedures adopted across the public sector. Secondly, the process and culture behind tendering in central government and local government are nevertheless distinctly different and have a bearing on the response of consultants.

From the late 1980s, central government has implemented a two-stage strategy, which aims to improve the efficiency in the use of property in two ways (Figures 13.1–13.3).

Firstly, responsibility for managing property has gradually been devolved from a central department, originally the Property Services Agency (PSA) and later Property Holdings, to the service level or departments. Service managers are therefore being given responsibility for decisions about how property resources are applied to delivering the service within their budget.

Secondly, government property services have been subjected to 'market testing' and the technical functions have been largely transferred to the private sector (Figure 13.2). The *Government's Guide to Market Testing* (published by the Efficiency Unit in 1993) states the aim of market testing as 'to promote fair and open competition so that Departments and Agencies can achieve best value for money for the customer and the tax payer', and identifies one of the key advantages as the creation of 'an explicit customer/supplier relationship'. This split is being reflected in the way property work is commissioned from the consultancy market, as discussed below.

The difference for local government is that Compulsory Competitive Tendering (CCT) for 'white collar' services is being imposed under the Local Government Act 1988 (Figure 13.3). The current legislation does allow councils to invite a bid from in-house staff who had been undertaking the work previously.

A number of local authorities have exposed all or part of their property function to market competition in advance of CCT, because they perceived advantages for the organization and staff in entering the market early on a voluntary basis. Five Council managers and three of their consultants, who had been party to voluntary competitive tenders (VCT), were interviewed for the study. They all felt that willingness on the part of the Council and its staff were key to building a relationship of trust within which the contract could be made to operate.

Under CCT, at least 65%, by value, of 'defined activities' within the 'construction-related services' function must be exposed to competition

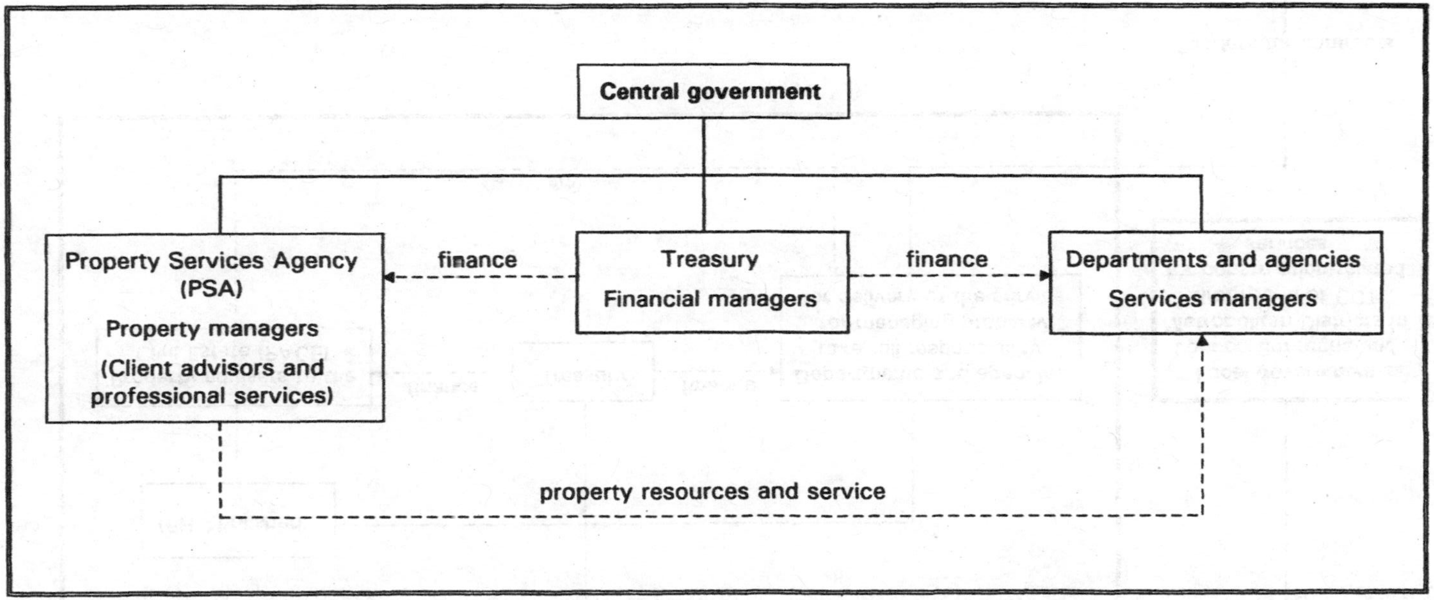

Fig. 13.1 Government Property Management; the late 1980s.

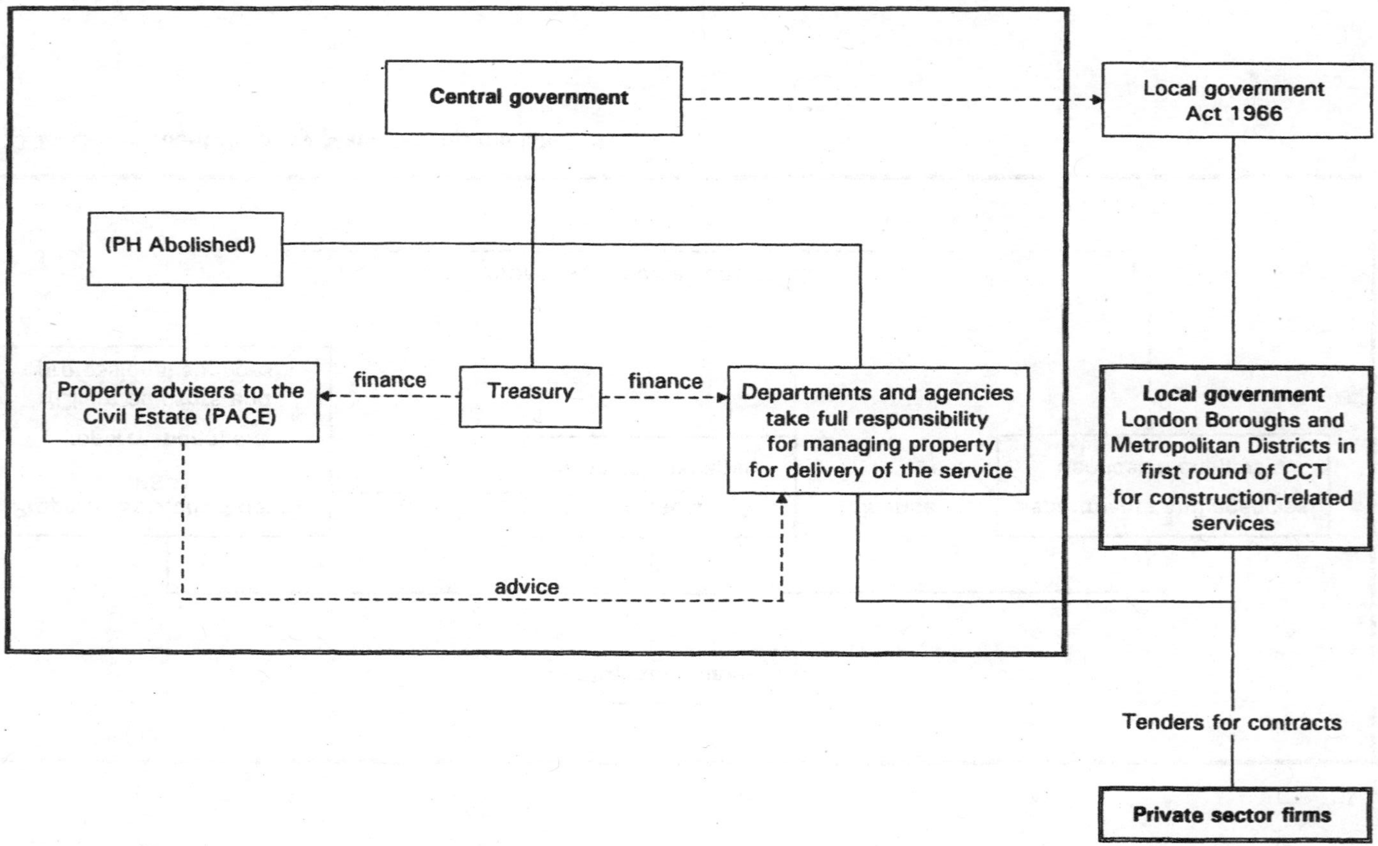

Fig. 13.2 Government Property Management; from 1 April 1990.

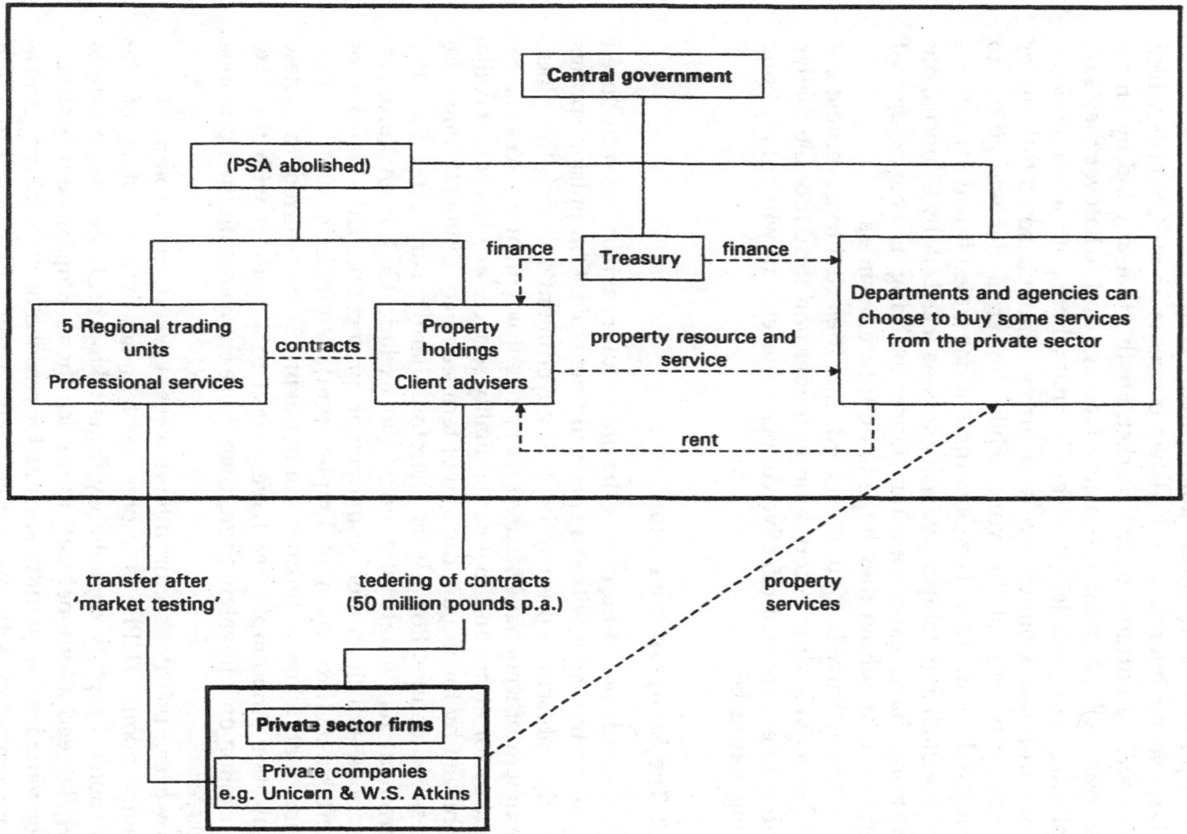

Fig. 13.3 Government Property Management; from 1 April 1996.

and 35% may be retained in-house. It is for each authority to decide which functions to retain. The signs are that information on property assets and a capability to deal with sensitive or strategic property issues will be a priority for retention among many councils.

Private sector interviewees had two concerns about Councils which are not willing entrants to the market. Firstly, that any bid by an in-house team will be given an unfair advantage and 'whatever they say, the playing field is not level'. Secondly, that where part of the service is retained in-house Councils are structuring contracts so as to keep for themselves the best of the work, while the remainder will go out to externalized contractors. The Government has stated it will use powers to combat such anti-competitive behaviour as it calls it, by Councils, but most firms have expressed reluctance to bring a charge against authorities with whom they hope to do future business.

The Labour Party has put on record, that if elected to government it would effectively drop compulsion to tender and introduce alternative methods to ensure cost effective council services, but without affecting existing contracts.

13.3.2 The European dimension

The UK stands out in Europe as the only country which has introduced compulsion to tender within a specified time-scale. Compulsion springs from the present Conservative-led government's belief that local government officers have secured terms of employment in excess of the private sector norm, and that better quality services at a lower cost could be achieved by forcing a reduction of terms and conditions through the process of competition. It is therefore hardly surprising that the government has fought strenuously to exclude CCT from European regulations which confer employment protection. The Transfer of Undertakings (Protection of Employment) Regulations 1981 (TUPE), provide that when a business undertaking is transferred to a new organization, the employees have a right to a contract with the new organization on equivalent terms than those enjoyed with their previous employer.

However, private sector interviewees viewed the prolonged uncertainty about TUPE as more problematic than the fact of the regulations. Property consultancy firms taking on large new contracts generally need additional staff to service them. Employment contracts can be varied subsequently, such as for reasons of internal reorganization unconnected with the transfer. For reasons like this, the TUPE regulations are therefore not necessarily a barrier to bidding. Several public and private sector interviewees shared the view expressed by a local authority manager that the two main areas of concern from firms at the point of submitting a bid were:

- the level of guaranteed income and does it cover the staff costs; and
- did they think that they were taking over a business or did they think they were taking over a disparate group of people?

In relation to European legislation, public and private sector managers were rather more concerned with the effect of the Services Directives, which require defined public contracts above a specified value threshold to be advertised across Europe after 1993 (contained in the Services Directive 92/50). Private sector interviewees reported a daily ritual of scanning the *Official Journal of the European Union* for notices of potential contracts, while government managers had the problem of properly evaluating the large number of responses.

13.3.3 The choices and challenges for property service managers

The attitudinal questionnaire sought to elicit the factors that were most likely to encourage firms to bid for public sector property work. Three issues emerged as most important and were locality of the work, followed by the type or mix of work, and thirdly the prospect of success in winning the tender. Locality and mix of work were two factors which also stood out in a Department of the Environment-commissioned study [BMRB International (1995)] of 'blue collar' local authority CCT contracts.

Prospect of success

This was very important to surveying firms owing to the cost of mounting a bid, for example, in one case 'whether it's worth spending £100 000–200 000 worth of senior staff time' mounting a bid against eight to ten other companies. Clarity of the client's specification, selection criteria and tender evaluation procedures are therefore very important in helping firms assess their own prospect of success at an early stage, and so avoid bidding for contracts they are clearly not qualified to fulfil.

Price and quality

In relation to tender selection criteria, both public and private sector interviewees identified these as the issues they felt most separated the basis on which public and private sector organizations select consultants or contractors. A number of interviewees shared the view that 'public sector clients have got their eyes over their shoulder for the auditor and they tend to go for the lowest price'. However, government managers interviewed later in the study expressed concern at the level of fee-cutting taking place in the market and said that they were prepared to justify why the lowest price should not be accepted if they felt quality could be compromised.

Quality was identified as being particularly difficult to judge for a service which involves the giving of professional advice rather than having a physical outcome which can be visually inspected. A local authority manager took the view that registration under BS 5750 would not

> . . . mean that I'm going to be satisfied with what they do, because in essence I'm going to strike a relationship with a group of people and we, as a team, are going to try and achieve something.

Several government departments were exploring methods to measure quality during the contract, for example through the use of benchmarking and auditing the work of surveyors. Performance-based contracts could also provide more scope for the consultant or contractor to exhibit flair and creativity, but prescriptive contracts were used by some government departments as a manual to guide their own in-house managers through unfamiliar territory.

Client's role

Property consultants also recognized the increasing importance of this and expressed greater job satisfaction working for a client who was strong, challenging and adequately resourced. Preliminary analysis of data indicates at least 50% of all respondents took the view that professional qualification was less important for the in-house manager than possessing:

- good contract management and procurement skills;
- a knowledge of property; and
- access to the decision-making level in the public sector body.

One consultant also commented,

> . . . on jobs where there is a single focal point with the client we have a better chance to build up a good working relationship.

Two research studies have identified firstly a general lack of proactive property management by major property occupiers, and secondly an emerging market for strategic property advice or property resource management. A number of government departments were using the tendering and private sector expertise to help them deal with both issues. As one government manager explained:

> We have taken the view that in the climate in which we operate that we do not need a whole bunch of in-house professionals and . . . therefore what we need most is **an intelligent client capability** . . . a national property adviser will provide two main roles for us. One is they will be our strategic partner and adviser

> . . . and secondly they will provide us with an audit function of the local estate management surveyors.

Interest in developing the challenge of the client adviser role appeared to increase during the CEM project in 1995, during which period the property market worsened. However, the RICS study suggested that general practice surveyors were in danger of losing out to other disciplines, through having established a more widely accepted reputation as 'deal makers' at the expense of developing management consultancy skills.

The choice of bidding for 'client adviser' contracts is an important one for firms, because it could preclude them from bidding for estate management commissions from the same organization. Larger regional firms in particular must therefore weigh up which path is likely to be most profitable for them. Several interviewees from smaller and medium-sized general practice firms were concerned about their prospect of success on two counts. Firstly, that despite having developed expertise in the client adviser role, there were signs that a lack of regional presence may preclude smaller firms from future appointments. Secondly, there was concern that they would be too small to take on some of the larger estates management contracts being brought to the market by both central and local government. Small firms were also more cautious about absorbing relatively large numbers of public sector staff, which they felt could swamp a currently successful and close-knit team.

For the **'contractor role'**, private sector firms found that having the right team leader in place from the start of the contract was very important to building a successful relationship with the client. An interview is often part of the tender selection process and it is important for the firm to field a team of people who will actually undertake the work. A managing partner who would not be directly involved in servicing the contract tended to ruin it for the staff involved at any interview. However, support by senior management for the public sector team is very important for the motivation of staff within the firm.

Failure in winning or running contracts provided some useful lessons, as one partner explained, '. . . we decided we had got to refine our presentation skills a bit and I think it stimulated us as a team'. Another interviewee confessed, 'we were using landlord managers . . ., and that was wrong. An occupier is very different to a landlord. You need different people'. However, dedicating a whole team to one fixed-term contract could have an adverse effect on staff morale. There is therefore a balance to be struck between servicing the needs of the client for dedicated attention and maintaining quality performance from the team.

Government managers saw signs of the market slowly adapting to provide the 'one stop shop' or 'seamless property service' that they are seeking. Several multi-disciplinary and facilities management firms had won contracts involving general practice work. Some had taken general practice surveyors into their organization, while others sub-contracted this element which they viewed as ancillary to their main business. Recent research by Ernst & Young on the private sector response to CCT identified a growth in the market towards facilities management, but expressed doubt that the market is sufficiently developed to provide facilities management arrangements to the smaller contracts, for example, those let by some district councils.

13.4 CONCLUSIONS

The success of the tendering process is dependent on the interaction between the parties, on establishing the right chemistry and a shared view about what needs to be done. As Sir John Harvey-Jones said, 'you cannot force good business approaches through organizational means. It's a cultural thing'.

The first challenge facing private sector firms is therefore to understand the requirements of individual public sector client organizations. The firm must then select those opportunities which it is qualified to service. One consultant identified that the performance of professional service providers actually reflects upon the client, but is also dependent on the client's ability to communicate what is required. Communication is therefore vital to bridging the gap between the public and private sector cultures. A company director explained:

> . . . we have had periods in the past where we couldn't do anything right for the council, but what was fundamentally wrong was the inability to communicate properly what we were doing.

Acknowledgements

This chapter was based on a paper from Gaye Pottinger from the College of Estate Management, who would like to thank the members of the Advisory Panel who have guided the research project and provided valuable advice in their own time. The panel comprised chartered surveyors Alan Darg, James Grierson (Donaldsons), Laurence Johnstone (Rogers Chapman), David Oram (Royal Borough of Windsor and Maidenhead). Thanks are also due to all those who have been interviewed and completed questionnaires. The project was independently funded by the College of Estate Management and overseen by Dr Tim Dixon, Research Manager.

13.5 REFERENCES AND FURTHER READING

Avis, M., Gibson, V. and Watts, J. (1989) *Managing Operational Property Assets*, University of Reading Department of Land Management and Development.

Bennett, M. and Whitehead, A. (1994) *TUPE – the EU's revenge on the 'Iron Lady'?* Cardiff Business School, Employment Research Unit, 1994 Annual Conference.

BMRB International Ltd (1995) *CCT: The Private Sector View*, Department of the Environment Local Government Research Programme.

Efficiency Unit, Office of Public Service and Science (1993) *The Government's Guide to Market Testing*, HMSO, London.

Ernst & Young (1995) *Analysis of Local Authority CCT Markets*, Department of the Environment Local Government Research Programme.

Meall, L. (1994) Retired but not retiring. *Accountancy*, June 1994, 64–6.

Pottinger, G. (1995) *Changes to Public Sector Procurement of Property Advice – Implications for Private Sector Firms*, The College of Estate Management.

The Royal Institution of Chartered Surveyors (1995) *The Chartered Surveyor as Management Consultant – An Emerging Market*, RICS, London.

Transfer of Undertakings (Protection of Employment) Regulations 1981 (TUPE); as introduced by the European Acquired Rights Directive (ARD).

Works & Supplies Remedies Directive adopted by the UK in 1989, amended by Services Directive 92/50 extends to UK contracts for services from July 1993.

CHAPTER 14

Marketing methods; client referrals

14.1 INTRODUCTION

This chapter looks at the importance of recommendation and referral to the marketing of building surveying services by practices throughout the UK. A brief review of the relevant literature is followed by a description of the way in which a research study into this topic was carried out. After a presentation of the main results, we will conclude with a discussion of how these results can assist all construction and property professionals in maintaining a strong source of new instructions.

14.2 THE APPOINTMENT PROCESS

How does a client select a particular practice to perform a professional service? Before the advent of fee competition, when all professional services were based on scale charges, and each firm provided virtually the same core service, the major form of obtaining commissions was based on attracting the prospective client to your firm by promotional techniques other than traditional advertising. This is classic marketing, in that an awareness of the service is created in the client, who will have already demonstrated a need.

In Part Four, we discuss in detail the nature of professional services in the construction and property industries. The professional practice should have a symbiotic working relationship with the client organization (or individual as the case may be). This means that marketing by impersonal means would probably not be as effective as that carried out by personal contact. There are five groups of methods of communicating the firm's message to potential clients, which the clients in turn will use as external sources of information to enable them to come to an objective decision as to which service provider to appoint.

1. **Impersonal advocate**, e.g. mass media advertising and sales promotion. On the face of it, this is the most expensive form of 'market

communications' but reaches a very large, if unselective, audience. Therefore the cost per person reached is low but is generally untargetted.

2. **Impersonal independent**, e.g. consumer reports and reports from the professional institutions. These are often considered to be very reliable because of their independence, but will always be generic in nature and unlikely to be specific to a particular practice.
3. **Personal advocate**, e.g. salespeople. The partners and/or directors have always been the sales force of the traditional professional practice, supplemented often by 'Consultants to the Practice' and by the 'referrers' who may be previously satisfied clients.
4. **Personal independent**, e.g. friends or colleagues. As and when business opportunities arise, friends who have no financial or other interest in the Practice may refer potential clients. The independence will hopefully ensure that the source is considered reliable.
5. **Direct observation and experience**.

The fundamental nature of services, as opposed to physical goods, is that they are intangible, that production and consumption occur simultaneously, and that 'sampling' of the service is very difficult, if not impossible. Therefore, direct observation and experience are unlikely to be effective for new clients.

The use of advertising and salespeople is unusual (although obviously becoming more prevalent) by professional service firms, mainly because of the regulations of the professional institutions which forbade such activity. The subjective nature of service quality means that few impersonal independent sources of information exist. Thus, the buyer of professional services is left with one major source of effective communications to potential clients – that of personal contacts.

This view suggests that clients or purchasers of services seek and rely more on information from personal sources than from non-personal sources when evaluating services before purchase. Also, the intangible and non-standardized nature of services leads to a high level of perceived risk and uncertainty on behalf of the client. This therefore brings an added dimension, in that the client is not only assessing the professional service being considered, but also the credibility of the referral source recommending that service.

Information about service providers obtained from a personal, independent source is generally defined as a 'referral', and it is this which we will concentrate on in this chapter. Let us start by looking at the nature of the term 'professional' and 'professional services', although this will be further examined in more detail in Part Four.

A professional has been defined as being 'an authority on his subject, as a body of knowledge, and an expert on its application to the solution of particular problems presented to him by the client'. This means that the client is therefore placed in a subordinate position in his relationship with the professional, and must rely on the judgement of his professional service provider. Depending on the client's experience of this relationship, this situation provides the basis for the operation of referral systems.

Of course, clients do use other non-personal sources of information to appoint professional advisers. Indeed, much of the literature about marketing of professional services is concerned with advertising, communications and the use of brochures. However, professional service providers have long practised what is probably the most effective form of marketing even allowing for the restrictions of the various professional bodies – the cultivation of influential contacts to encourage referrals. Few professional firms advertise, even though they are now free to do so, and this culture still exists with both professional firms and clients tending to resist the use of standard advertising techniques.

14.3 HOW IMPORTANT IS RECOMMENDATION AND REFERRAL?

According to Aubrey Wilson, in the qualifying professions estimates of all new business from referrals range from 80–100%. There can be few business activities where the overwhelming value of business is acquired through a single marketing tool and yet so little knowledge exists of that tool or so few attempts have been made to establish whether it is nevertheless capable of further exploitation.

Nor it seems has this view been overtaken by time. In 1992 Herriot remarked that intercorporate referrals have received scant attention in marketing literature while in a recent and otherwise thorough description of the history of services marketing the subject is not even considered worthy of mention. Connor and Davidson suggest that cultivating referrals is the backbone of a successful practice, yet too many professionals regard this as a burdensome or worse, unimportant task.

Herriot goes on to define a referral channel as having a one-way flow of referrals but a referral system as having two or more directional flows. He notes that channels are usually asymmetrical in that more work usually flows in one way than the other. He points out also that while it is obvious that the firm referred is dependent upon the referrer, the referrer is also dependent upon the firm providing satisfactory service. If the service is unsatisfactory the referrer will lose some respectability with the person to whom the recommendation is made. Generally, therefore, members of a referral system are dependent upon one another.

Connor and Davidson define two classes of referrer, satisfied or enthusiastic existing and previous clients and non-client influentials who provide leads or will mention professional firms to their clients and others. They recommend taking positive steps to identify and analyse referral sources. These steps are similar to those suggested by Wilson which are identifying sources of referrals, acknowledging and reciprocating referrals, keeping the referrer informed of progress and establishing the reasons for referrals.

Recent research undertaken by Building Surveying staff at the University of Salford as a prelude to investigating just how important client referrals are to this particular profession started with the following hypothesis:

> That more than 80% of instructions from new clients of building surveying firms are gained through recommendation and referral.

In addition to measuring the proportion of new instructions emanating from referrals the study was designed to ascertain the source of these referrals and the relative importance of factors considered by clients when appointing building surveyors.

14.4 RESEARCH PROJECT

The main research instruments were postal questionnaires sent to building surveying firms and their clients throughout the UK, but in order to take a closer look at the working of the referral system case studies of actual jobs of the firms were also carried out. The two questionnaires were similar in format and employed 5-point unbalanced Likert-type attitude scales to assess the importance of 14 appointment process factors. These factors were taken from a literature review, from a study of general practice surveying firms undertaken by the department, from the writer's own professional experience, and from a pilot study. Respondents were asked to indicate their level of agreement (from 'disagree' to 'very strongly agree') to the assertion that each factor was important to the appointment process. The questionnaire asked firms to estimate the proportion of instructions from new clients and for these new clients to estimate the proportion appointed as a result of five methods. Clients were asked to indicate which method they had used to appoint their current consultants. In total, 169 (67.6%) completed questionnaires were received from firms and 126 (46.3%) from clients. These responses are well above average for a single mailing survey.

The 72 case studies over 12 firms investigated the method of appointment for each job (selected at random from a chronological list of the firm's most recent 100 jobs) and for the cases where clients were new and where recommendation and referral had been the method of appointment, the referral chain was traced back and a note made of the media used to make referrals.

14.4.1 Results

Of the clients responding to the questionnaire survey, public body clients (government departments, health authorities, universities, county and district councils) represented 45% of the sample, corporate clients made up 30% and the remaining 25% consisted of small firms, private individuals, housing associations and churches. There was therefore a wide range of client type in the sample. The ages of the firms and their sizes (measured in terms of the number of partners or directors) were typical of building surveying which is a relatively young profession having a few large firms and very many small firms. Some 80% of the firms' questionnaires were completed by either a partner or director. A sampling frame of 500 firms was identified as providing building surveying services. The sample studied is almost exactly one-third of this population and since the sample has been shown to possess the characteristics of the population it is likely that the study will faithfully reflect the views of the population.

The results of the analysis of, what is admittedly a small number of case studies, supported the original hypothesis since 85% of new clients were introduced by recommendation and referral. The maximum number of links in the referral chain was 4 and the mean number of links was 1.5. The chain is therefore fairly short and informal means of communication predominate. Small, inexperienced clients rely on the referral system much more than do larger, more sophisticated clients. In the questionnaire study the firms' mean for the proportion of instructions from new clients was only 35% and of these 61% were introduced through recommendation and referral. The sources of instructions are illustrated in Figure 14.1.

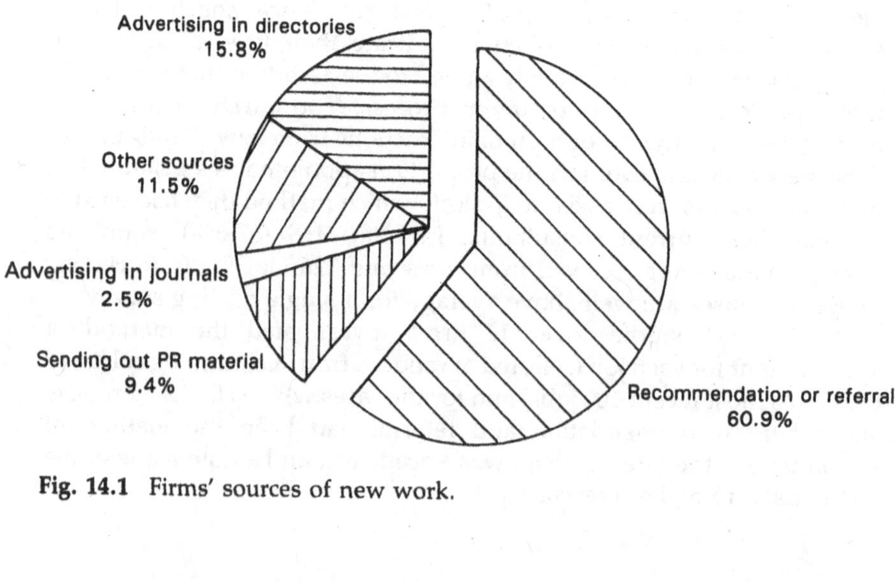

Fig. 14.1 Firms' sources of new work.

Overall, clients said that of their current consultants 50% were appointed through recommendation and referral. Neither of these figures is as high as hypothesized but recommendation and referral is by far the most important method of appointment by new clients. The two main referrers are existing clients (43% according to firms and 26% by clients) and other professionals (38% firms and 66% clients). Estate agents are both parties' most important professional referrer, accounting for over one-third of this category of referrer.

A full picture of these professional referrers as stated by the firms is illustrated in Figure 14.2.

Figure 14.3 gives a profile analysis illustrating the relative importance of the appointment process factors as assessed by firms and clients. It will be seen that the attitudes of both parties are approximately parallel but at a probability level of 0.01 there are six of the 14 factors where the responses differ significantly. The factor for which the greatest difference was recorded is of particular relevance to this research and is **recommendation and referral by a third party**. The ratio of the mean scores is 1.22, which is identical to the figure achieved by dividing 61% by 50%. Thus overall, each party's attitude to this subject is a faithful reflection of their stated behaviour. This result strongly supports the attitude theory held by marketers that cognitive, affective and behavioural components of attitudes tend to remain in balance and therefore tends to support the validity of the data provided by this part of the study.

The ranking of the appointment factors, illustrated in Figure 14.3, is

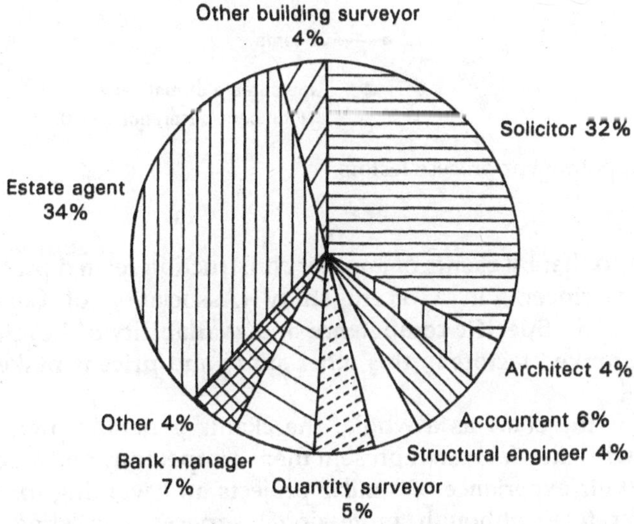

Fig. 14.2 Firms' professional recommenders or referrers.

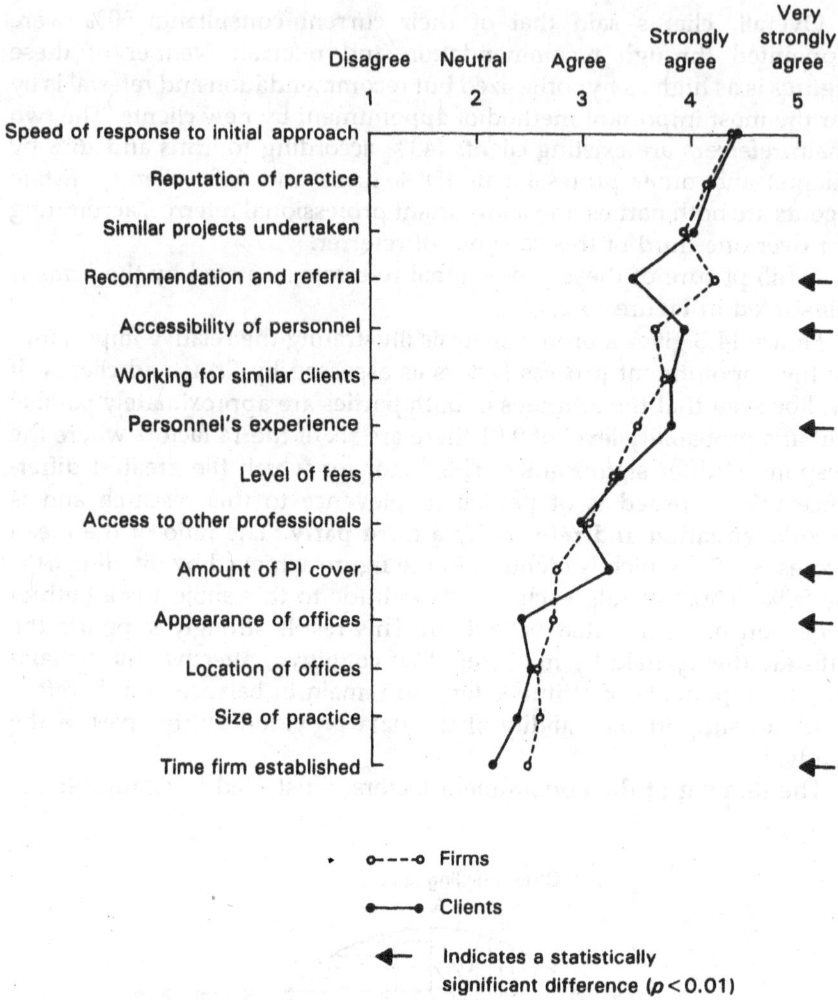

Fig. 14.3 Appointment process factors.

very similar to that by clients of another construction related profession, consulting engineers, in a recent study (Association of Consulting Engineers, 1995). **Specific competence** and **availability of key staff** are the highest ranked factors in this other study and **price** is ranked only fifth.

Looking at the chart as a whole, the clients generally prefer those companies who they feel can represent them competently and efficiently, shown by their experience of similar projects and working for similar clients. Therefore, although professional services marketing makes much of the fact that it is *ad hoc* by nature, and that each commission is tailored to the particular client, the clients nonetheless wish to see evidence of similar work undertaken. The 'Speed of response' and 'Accessibility' criteria show that clients wish their professional advisers to be efficient and accessible when working on the client's behalf, possibly mirroring the efficiency and speed of response with which the client organization treats (or would like to treat) its own customers.

The fee levels are given as mid-range in importance, illustrating that service is a higher priority than cost, which itself has some interesting ramifications for the future of Compulsory Competitive Tendering discussed earlier in Part One.

The factors which the clients (and firms) do not see as important can also be grouped together, as the high profile symbols of corporate life, such as appearance and location of offices and when the firm was established. A great deal of resources are put into these factors as promotional tools, with apparently limited impact, according to this study. However, as they also broadly correspond to Herzberg's Hygiene factors, then perhaps the benefits may be seen in improved staff satisfaction and motivation.

14.4.2 Recommendations

How then can the results of this study assist building surveyors (and possibly other professionals) in maintaining a strong source of new instructions? The following points can be drawn out as recommendations.

1. **Ensure that existing clients are satisfied**, since this study suggests that overall, only 35% of new instructions are from new clients, the emphasis on a Practice's marketing strategy should therefore be on satisfying existing clients to ensure that they are retained. Existing clients are always your best clients.
2. **Improve service quality**, not only to satisfy existing clients but also to ensure that satisfied clients and/or their professional advisers recommend the firm to other potential clients. The study suggests that between 50% and 60% of new clients are recommended or referred and that between 80% and 90% of these referred clients are referred by existing clients or by other professionals.
3. **Take positive steps to encourage referrals**. By identifying sources of referrals, and acknowledging and reciprocating any referrals from other sources, the referrer can be kept informed of progress and the reasons for referrals established.
4. **Undertake other marketing activities**, especially where the firm is young, expanding, or competing with more established Practices for a limited amount of work. Given the results of this study, under normal other circumstances, other forms of marketing are only likely to influence at most 17½% of new instructions.

5. **Be aware** of the relative significance of factors clients consider important when appointing professionals. Figure 14.3 suggests that more emphasis is placed upon the speed with which the professional responds to the client's initial approach, the reputation of the Practice and whether it has undertaken similar work than such things as the level of fees quoted and the size and age of the Practice.

This research has therefore provided strong support for the observations of many experienced professionals: that the best marketing tool available is to ensure that existing clients are satisfied by the service provided. These clients will not only remain clients themselves but also act as ambassadors for the firm.

14.5 REFERENCES AND FURTHER READING

Association of Consulting Engineers (1995) *The Fisher Report: The Role of the Consulting Engineer Now and in the Future: The Client's Perception*, Association of Consulting Engineers, London.

Banks, J. and Barrett, P. (1992) A Synthesis of Clients' Criteria for the Assessment of Professional Firms, in *Proceedings of the Architectural Management International Symposium* (ed. M.P. Nicholson), University of Nottingham.

Brown S., Fisk, R. and Bitner, M. (1994) The development and emergence of services marketing thought. *International Journal of Service Industry Management*, **5**(1), 21–48.

Connor, R.A. and Davidson, J.P. (1985) *Marketing Your Consulting and Professional Services*, John Wiley & Sons, Chichester.

Donnelly, J.H. and George, W.R. (1981) *Marketing of Services*, American Marketing Association, Chicago, Illinois, USA.

Dornstein, M. (1977) Some imperfections in the market exchanges for professional and executive services. *The American Journal of Economics and Sociology*, **36**, 113–28.

Harris, B.F. (1981) Strategies for Marketing Professional Services, Current Status and Research Directions, in *Marketing of Services* (J.H. Donnelly and W.R. George), American Marketing Association, Chicago, Illinois, USA.

Herriot, S.R. (1992) Identifying and developing referral channels. *Management Decision*, 30(1), 4–9.

Tull, D.S. and Hawkins, D.I. (1976) *Marketing Research*, Macmillan Publishing, New York.

Wheiler, K. (1987) Referrals between professional services providers. *Industrial Marketing Management*, **16**, 191–200.

Wilson, A. (1984) *Practice Development for Professional Firms*, McGraw-Hill, Maidenhead.

Wittreich, W.J. (1966) How to Buy/Sell Professional Services. *Harvard Business Review*, March/April, Cambridge, Massachusetts, USA.

Zeithaml, V.A. (1981) How consumer evaluation processes differ between goods and services, in *Marketing of Services* (J.H. Donnelly and W.R. George), American Marketing Association, Chicago, Illinois, USA.

CHAPTER 15

Marketing methods; relationship marketing

15.1 INTRODUCTION

So far, in this book, we have talked about the more traditional forms of marketing as applied to professional services. We have mentioned the traditional forms of market analysis as applied to professional firms in the property industries and made specific references to the service/market matrix (Chapter 12) and client referrals (Chapter 14). Both of these are valid in their own right, but can also be seen as part of a wider set of tools for the promotion of organizations who depend for their prosperity on developing a relationship, both between them and their clients, and between the organization and its own staff. Uniquely, in professional service organizations, it is the staff of the firm (i.e. the individual architects, surveyors, engineers, etc.) who provide the service, react to clients' demands and earn the fees. The role of management must necessarily be one of facilitating this service by ensuring that adequate resources are available as well as providing professional development opportunities in these constantly changing times. This is what Charles Handy describes as the 'upside-down pyramid'. Customer (or client) satisfaction is determined by the relationship developed between the individual professional and the client; as far as the client is concerned, the individual professional is the firm and those charged with marketing the firm's services must recognize this if they are to be successful. There is therefore a strong relationship between customer satisfaction and marketing.

We also look, in Part Four, at the question of quality of service and whether formal systems are necessary for the client to obtain an adequate service level. Inadequate service quality soon becomes clear in an increasingly competitive marketplace, with firms giving poor quality finding that their commissions are drying up fairly quickly. There is therefore a strong relationship between quality of service and marketing. The third link is fairly obvious, that there is a strong link between customer satisfaction and quality of service.

We now have the basis of what relationship marketing is all about, the triangular connections between marketing, customer service and quality.

15.1.1 Marketing

In Chapter 11, we looked at the nature of marketing and applied the traditional tools to the professions in the land, construction and property industries. We saw that a company needed to develop a differential advantage, or USP, in order for it to be attractive to clients and put a quasi-monopoly price on its services. It also must 'position' itself in the marketplace. To do this, it uses the four 'Ps' of Product, Price, Promotion and Place, which were discussed in some detail. However, this model assumes that the firm – and remember that we are talking about firms providing professional services – acts as a 'supplier' to client organizations. By this, we mean the professional firm has products and services which can be bought off the shelf by clients, who have no interest in how the firm organises itself as long as they are satisfied with the final product.

There is a valid argument that this has never been the case. Traditionally, professional service firms acted in a *locum tenens* capacity for their clients (in fact, barristers still do act in this way, hence the term 'counsel'). We could say that since the relaxation of the rules of incorporation and advertising and abolition of the scale fees, professional firms became much more a supplier of services, but we are also now seeing professional firms deliberately investigating the culture of their clients, almost all of whom are corporate. Professional firms now put a great deal of effort into how their clients make decisions, who makes the decisions, what time-scales are important to them, and where are the critical points of information flow.

Where is all this leading? If professional service firms are to be successful in developing corporate relationships with their clients, then two further Ps must be considered and added to the marketing mix – People and Processes.

People

A professional service firm is the people it employs. Not just the Partners or Directors, or even the technical and professional staff, but everyone. The traditional four Ps include people issues in the promotion area, and we have discussed it previously when considering what creates image. We have also looked in Chapter 9 about the most effective leadership styles for professional people. Bringing all these together, plus the motivation of professional people and how they react when working in groups, we surely have enough material for a new item in the marketing mix.

Processes

This also has a valid claim for a separate role within the marketing mix, especially when also considering quality management and increasingly demanding clients. Process management involves the procedures, schedules, mechanisms, activities and routines by which the service is delivered. The dramatic increase in information available to almost anybody now who has the appropriate technology, taken with the increase in competition for services traditionally offered by architects, surveyors, etc. means that processes must not only be quicker but offer a value added to the client not available elsewhere.

An example of this is in the financial services industry, where the lead time for the introduction of new products can be as short as 6 months. This means that the information required for market research and product development must not only be instantly accessible, but in a form which can be used effectively by the sales team. The product life-cycle, mentioned by Peter Lansley in the introduction to this Part, has shortened considerably due to the introduction of efficient competition, requiring that the processes of sales are shorter and more cost efficient, using techniques such as automatic telephone enquiry systems.

While the people element of professional services is important, poor process performance may result in incorrect or untimely advice, with the obvious repercussions for the reputation of the firm.

15.1.2 Quality

It does seem strange to think of quality as a marketing tool, especially as we define quality, not in its Rolls-Royce sense, but in its definition as 'fitness for purpose' or the more subtly titled 'customer-perceived value'.

In discussing quality within the professional service sector, in the context of this area of marketing, or building a symbiotic relationship between firm and client, there are five basic quality dimensions which would impact on service delivery:

1. **Reliability**: does the firm have the ability to perform the promised service with dependability, accuracy and consistency? Obviously, all professional firms will say yes to this question, but given the points made above, rate of change in today's economic climate, the downsizing of firms, and increasing staff mobility, what strategy does the firm have for maintaining reliable services in the long term?
2. **Responsiveness**: this links with the points made above about the shortening product life-cycle and clients' requirements for instant responses. Building a relationship with a client often means virtually being part of their organization and understanding their constraints

and requirements. Do we have the flexibility to be instantly responsive?
3. **Assurance**: are our professional and support staff assured of their professional competence and their ability to satisfy clients' requirements? Have they the ability to inspire trust and confidence?
4. **Empathy**: we have mentioned before that professional service firms developed as *locum tenens* of their clients, and that relationship marketing is a way for the new commercially oriented practices to rediscover and redefine that linkage. Central to that linkage must be the concept of empathy, with the caring, individualized attention within the framework of the client's culture.
5. **Tangibles**: we have already mentioned in Chapter 11 that image is developed by the physical facilities, equipment, staff appearance, etc. of the firm. In order to achieve a quality service, these 'Hygiene factors' (as Frederick Herzberg called them) should also be appropriate and convey both functional and symbolic meaning.

15.1.3 Customer service

Customer service is a relatively easy concept to define in terms of manufacturers of physical products, or even providers of services such as dentists, where a need arises (toothache), the service is delivered, if the symptom goes away then the customer is satisfied. In both cases, the customer is more concerned with the satisfaction of a want or need than with the processes and procedures. Why the customer has gone to that particular dentist in the first place is another story, and is more related to the material in Chapter 14, referral marketing. An interesting example here, where processes contradict the image, is the television advertisement in the UK where a hairdresser offers a 'Lionel Blair' haircut (in fact, a spikey short back and sides) in total contrast to the high-quality decor of the salon. The customer exits to the nearest local hostelry to drink the advertised product with a sigh of relief at escaping the fate. In terms of our argument, the customer perceived a poor quality of service, would be highly unlikely to return to the provider, and additionally would be likely to negatively refer the salon to his friends and colleagues.

It is very difficult to define customer service, although everyone knows whether or not they are getting it, as seen in the example above. However, the following do contribute to effective and positive customer service:

- The processes and procedures required to accept and deliver customer orders or commissions or to follow-up on any complaints or queries.
- Timely and reliable delivery of and invoicing for the completed service in accordance with the customer's expectations and internal procedures.
- Appropriate following-up of service delivery.

Customer service, within the context of relationship marketing, is therefore concerned with the building of bonds, to ensure long-term relationships of mutual advantage. It is the holistic process which provides satisfaction both before, during and after the actual delivery of the service or commission. Quality customer service involves an understanding of the client purchasing process and showing how our services can provide additional value to their organization, including establishing whether or not this has occurred.

15.2 NATURE OF RELATIONSHIP MARKETING

Is it the bell that rings?
Is it the hammer that rings?
Or is it the meeting of the two that rings?

15.2.1 Customer orientation

Relationship marketing in terms of professional service firms has three major components or elements:

1. Employees and suppliers.
2. Clients and other purchasers of the service.
3. Non-purchasing but influential external firms or individuals.

Whereas the primary focus of traditional marketing is aimed at the external client – whether they are actual or potential purchasers of the service – the focus of relationship marketing is a much more holistic process and encompasses the internal customers which includes their input to the processes and procedures of the firm. All employees, and especially professionally qualified staff, have a contract with the firm which is both legal (contract of employment) and psychological (in terms of expectations on either side). Highly qualified and experienced professional staff have a high value in the marketplace, and in order to retain these people, traditional theory has stated that the firm must pay a 'loyalty bonus', which is usually the difference between what they can obtain elsewhere. These employees are therefore customers of the firm, in virtually the same way as external clients, and relationship marketing recognizes this as it is a fundamental tenet that satisfaction must exist on both sides for a full relationship to develop and mature.

Charles Handy, in *The Empty Raincoat* talks of modern companies being likened to a three-leaf clover, with one leaf being the core staff, another the temporary and part-time staff, and the third the sub-contractors and suppliers. He states that we had all better get used to this principle and not be frightened to be self-employed 'knowledge experts' plying our trades and professions through teleworking, hot-desking, hotelling and cottaging. True as this may be, and there is increasing evidence of industry taking up this opportunity for increased flexibility; for our purposes here, we should include suppliers and sub-contractors in our total relationship model. There is little point in promising increased customer service to clients if a critical part of the service is produced by sub-contractors over whom we have no direct control and are not aware of our commitments.

In the same way, client loyalty is built up not just by quality service provision, but by clients reinforcing their perceptions from independent sources. We have all made purchases from high street retailers, and been satisfied or otherwise with the quality of the purchase. If we have been satisfied we will tend to return to that shop for a repeat purchase when the time comes, or to see what else they have to offer which may satisfy a particular need or want. This perception is strongly reinforced when we talk to other customers who have also been strongly satisfied. Thus, customer loyalty is developed.

Relationship marketing is therefore concerned with 'influencers' and where these influencers are not clients themselves but are perceived as reputable independents, their opinions naturally increase in value. This point has been discussed in Chapter 14.

15.2.2 Developing employee loyalty

As the number and types of firms offering professional services in the construction and property industry increases both in terms of number and type (for example, Management consultants/Accountants are increasingly being appointed as the lead consultants on large property developments), there is strong competition in attracting a suitable number of motivated and trained employees. This is where relationship marketing takes a completely different view of the contract between employer and employee, and relates more to the psychological contract mentioned above.

Because of the rapid changes in technology, forms of procurement, legislation and standards, knowledge obtained at universities has a relatively short shelf-life compared with 20 or 30 years ago. Employing organizations in professions such as accountancy have recognized this and look for prospective employees with high level transferable and problem-solving skills, rather than the traditional technical or procedural skills. Many professional practices in the construction and property industries are now obtaining instructions in what they describe as 'management consultancy' rather than their traditional areas. Therefore, a different type of recruit is required, needing a reappraisal of the traditional recruitment policy.

How then do we encourage employee loyalty? To answer this, we must understand what motivates professional staff and how they expect to be treated. The latter point was discussed in Chapter 9 on leadership styles in professional firms, and relates to their expectations, interest in the job to be done, tolerance of ambiguity, past experience and other cultural factors. Motivation of professional staff is much more complex and can relate to the satisfaction of various needs, the expectancy of extrinsic or intrinsic rewards or their responsiveness to job related or environment-related signals.

An easy answer would just be to pay more in salary or performance-related bonus or in emoluments in kind. All the research into the motivation of professional people states that money does not act as a motivating agent; in other words, people will not necessarily work harder or better if they are paid more. They have a complex set of personal wants and needs to be satisfied, some of which are currently being addressed and some of which are on hold. Paying intelligent people well to do unchallenging jobs is a recipe for failure, since they will merely find more challenging or enjoyable pastimes which they have not been able to afford to do in the past.

A common thread through all of this is that professional people should be motivated by **challenge**, since responsiveness to whatever is requested of them is a fundamental principle of relationship marketing. It is therefore the responsibility of the firm to provide a challenging environment which can be created through the nature of the job, or other more structural techniques.

A topical theory in this area is the concept of the 'Learning Organization', which is a culture of continuous improvement and is a development of total quality management particularly applied to professional consultancies. The organization and its constituents is seen more as a biological organism than a static structure, with the ability to change shape and react instantly and flexibly to external (and internal) influences and demands. It does this by concentrating, not on power, authority and roles in the traditional sense, but on its staff, processes and procedures constantly learning and developing from information and data coming from the environment.

The choice that professional practices must make is whether to require staff to work harder at their assigned tasks (i.e. the traditional power/authority model) or to encourage them to participate in generating new and improved procedures for the firm, based on what information is reaching the firm from the environment.

15.2.3 Developing client loyalty

Traditional marketing theories put great importance on the retention of existing customers, and relationship marketing is no different, and if anything, puts an even greater emphasis on retaining existing clients.

The reasons for this are fairly obvious, existing clients and customers already know your services and are familiar with the people, processes and procedures of the firm. Retaining an existing client is far cheaper in terms of investment than attracting new clients and it is a fairly good measure of customer satisfaction if clients and customers are willing to come back for more, which is the point made with the high street retailer mentioned previously.

Expectations of service quality

Client loyalty is therefore a direct function of achieving their desired expectations, which itself is a measure of service quality since:

$$\text{Service quality} = \frac{\text{Perceived performance}}{\text{Desired expectation}} \times 100$$

Under this formulation, for professional services, anything less than 100% would be construed as a failure; there is very little margin for error in what is often investment advice to a client.

We can now see how the building up of relationships is vital to ensure that the clients' perceptions of the performance of the service that is offered is matched with their desired expectations at all stages during the commission, i.e. pre-delivery (briefing), delivery, and post-delivery (a stage which is too often ignored by professional practices, and which has given rise to service augmentations such as post-occupancy evaluation, as mentioned in Chapter 11).

The function of the professional firm is therefore to close any potential gap between the client's expectations and their perception of what the firm is offering. This creates four further potential 'gaps' between the professional service firm and client to be addressed during negotiations:

1. Is there a difference between the client expectations and the professional firm's perceptions of their expectations? This sounds as though we are merely playing with words, but is a common problem with communications – is what I thought you said what you actually meant to say?
2. Is there a difference between the firm's perceptions of their client's expectations and the service quality specifications? Are we promising something that we cannot deliver?
3. Is there a difference between the service quality specifications and the service which was actually delivered? This is a function of project management, to ensure that delivery is in accordance with the original specifications.
4. Is there a difference between the service delivery and how we have included that project in our marketing brochures or documents?

Addressing all of these issues during the delivery of a service to all clients should ensure that clients' expectations are consistently being met, resulting in satisfaction and loyalty.

Service performance auditing

Auditing of service during delivery is the way in which the professional firm can establish if there is a difference between the service quality specification and what is actually being delivered. A performance audit is defined as a 'systematic review of the performance of the service provided, in order to ensure that the client's requirements are being met', which is designed to ensure that there is a high probability of client satisfaction, when the service has been delivered. The effectiveness of service delivery can be affected by the following:

1. Behavioural aspects of people carrying out the service; for example, their attitude, motivation, reaction to status, role relationships, as well as any personal issues which may affect their work performance.
2. Techniques and technology, processes and procedures; do all the contributors understand the way that the work will be carried out?
3. Decision-making methods, which in itself is a process, and is likely to be developed to a sophisticated level within professional service organizations, although the service delivery (or project) manager must ensure that the process of decision making is efficiently designed for that particular purpose.
4. Temporary organization structures; the management of construction projects in particular require temporary organizations to be set up to deliver the service. Any organizational structure is a formal social system working towards an objective, and establishes structured relationships between individuals and groups of individuals. Poor design of these structures will create information bottlenecks, which is the major cause of problems in the construction industry.

15.2.4 Developing a relationship strategy

We have now seen that developing a relationship marketing ethos is as much about developing internal relationships with employees, subcontractors and suppliers, as with developing relationships with clients and potential clients. The firm must develop a 'shared vision' between all the people who input into the service and effective processes which communicate this vision. It is also, of course, about serving and retaining clients, which can be done most effectively by ensuring that customer satisfaction is high, thus developing loyalty and therefore repeat business.

In attempting to develop a strategy to help firms manage a change to relationship marketing, the American management consultants McKinsey & Co. developed what they called the 'Seven Ss' framework which is a powerful tool in planning and managing organizational change. The seven Ss are:

- Shared values
- Structure
- Strategy
- Systems
- Style
- Skills
- Staff

Shared values

All employees and inputters into the service delivery must share the same values and goals for total customer satisfaction based on building relationships, rather than deliver the goods, invoice and depart. This becomes even more crucial in the professional service industry where the advice given is often part of a greater decision-making process by the client, or when the client may not actually be buying your service but is merely collateral to their real purchase (an example here is that the client is not buying the quantity surveyor's service, they want a building in order to perform their own operations).

Structure

Simple organizational structures are usually the most effective when dealing with creativity and creative people, as discussed in Part Two of this book. Effective organizational structure has a close relationship with the appropriate culture for that type of operation, and for professional service firms offering independent, bespoke advice to clients, then a rigid formality in organizational structure is unlikely to be effective. There should be few formal divisions in the firm, marketing is the job of everybody, and staff and role rotation should be constant, in order to maintain a high degree of flexibility.

Strategy

A flexible organizational structure can only really work when the firm knows where it is going and how it intends to get there. Therefore the managers of the firm must put in place an integrated plan for the development of its marketing orientation, by defining its mission and its markets, specifying its objectives and show a commitment to its implementation by, for example, the Investors in People standards.

Systems

The systems, or processes, of the firm have been mentioned previously as an important aspect of relationship marketing. They must be designed to use high-quality, relevant and up-to-date information from reputable sources, with their output being useful to the professional staff giving advice to the clients. Customer and competitor intelligence reports are important aspects in today's business climate, with many professional practices having research or business development sections whose main duty is to develop this area.

Style

The style or culture of an organization can often reflect its commitment to clients as well as the commitment of top management to support the work of the professional staff through various forms of symbolic action or commitment of time. Communications must be open and internal politics minimized, otherwise the dedication of the professional staff will evaporate.

Skills

As we have mentioned above, the skills of professional staff within a firm committed to relationship marketing must of course be of a very high technical quality, but must also be flexible. This requires a high degree of what are called 'transferable' skills in man-management, etc.

Staff

All staff must be totally committed to the ethos of relationship building with clients, customers or potential clients and customers. Recruitment policies must reflect this commitment, as should the remuneration policies. A direct result of the upside-down principle mentioned previously, is that it is the individual staff who are the critical points in the firm, and their value should be reflected accordingly.

15.3 CONCLUSIONS

Relationship marketing is nothing really new – little in the world nowadays actually is new. There are few ideas or techniques mentioned in this chapter which have not been espoused before in marketing texts. What is new, however, is the way that they are put together. The world has changed considerably over the past 10 years, the construction and property industry certainly has changed, although the jury is still out on whether the change is for the better. The way that land is secured for development has changed, local authorities are 'commercialising' under CCT, architectural practices are losing commissions to design-and-build construction firms, two of the largest construction firms in the UK have actually stopped bidding for work in the early part of 1996, professional practices of surveyors are securing a greater proportion of their work as management consultants rather than their traditional areas of expertise. With this change must come a different approach to marketing and securing work, effectively a reinvention of the *locum tenens* philosophy of the professional practitioners of the last century, while avoiding the restrictive practices of their forefathers. This is the central core of relationship marketing, a reconfiguration of marketing techniques to serve the needs of the present and the future practitioner.

15.4 REFERENCES AND FURTHER READING

Christopher, M., Payne, A. and Ballantyne, D. (1991) *Relationship Marketing, Bringing Quality, Customer Service and Marketing Together*, Butterworth-Heinemann, Oxford.

Handy, C. (1994) *The Empty Raincoat*, Hutchinson, London.

Handy, C. (1991) *The Age of Unreason*, Business Books Ltd, London.

Katz, B. (1988) *How to Market Professional Services*, Gower, Aldershot.

Maister, D. (1993) *Managing the Professional Service Firm*, The Free Press, New York.

Peters, T.J. and Waterman, R.H.(1982) *In Search of Excellence*, HarperCollins, New York.

PART FOUR

Professional Ethics and Quality of Service

Introduction to Part Four: Managing quality and professional ethics

Ken Innes, retired Senior Partner of Scott Wilson Kirkpatrick Consulting Engineers

LINKS WITH FEE COMPETITION

Professional ethics are linked inextricably with the issues of fee competition discussed in Part Three. When compulsory fee competition was introduced in the UK for consulting engineers the Principal of an American firm wrote an article explaining his approach. He said that if appointed professionally to carry out a commission, it was his duty to use his professional skills solely in the interests of his client. When he was asked to bid competitively for an appointment he believed it was now a situation where he could use his professional skills in his own interest. He could look for weaknesses in the brief which could be exploited to his own benefit initially to win the job and subsequently to maximize his profits.

This clearly shows the way in which he saw ethics, quality and price, being linked together.

THE ASSOCIATION OF CONSULTING ENGINEERS' SURVEY

The Association of Consulting Engineers has been seriously concerned about the effect of fee competition on ethics and quality and because of this, it carried out a poll of its members. Some of the results were included in the Latham Report, *Constructing the Team*, and in fact were quoted by Sir Michael Latham in his introduction to Part Two. I would like to repeat a few.

Of those members who participated:

- 73% give less consideration to design alternatives;
- 31% give less consideration to checking and reviewing designs;
- 40% consider that the risks of design errors occurring are higher;
- 74% admit that they are producing simpler designs to minimize the commitment of resources to a task;
- 60% consider that capital costs of construction and operation are higher as a result of the previous point;
- 84% assess the number of claims for additional fees to be higher;
- 33% thought the frequency of problems on site are higher;
- 49% said the frequency of visits to site are lower;
- 29% said they pay less attention to environmental concerns;
- 12% pay less attention to health and safety both in design and on the site;
- 67% resist client changes to designs;
- 69% see less trust between client and consulting engineer;
- 79% are spending less resources on training graduates and technicians;
- 77% are spending less resources on Continuing Professional Development training and courses;
- 75% are spending less time on the writing of professional papers;
- 56% are devoting less time to professional activities;
- 94% bid low to maintain the cash flow or (on occasion) to test the market;
- 35% bid low with the intention of doing less than in the enquiry; and
- 61% bid low with the intention of making up fees with claim for variations.

LAWYERS

Looking to another profession, the article in *The Times* of the 24th August, 1995 entitled 'Solicitors seek end to cheap conveyance' reported the Law Society's concern about cut fees. Concern was expressed about the shoddy service the public is receiving from firms determined to cut fees. The Deputy President is quoted: 'What has happened is crazy. . . . fees have dropped to a level which is almost suicidal and dangerous, because the job can't be done properly for the rate and then there are claims for negligence'.

THE NEED FOR PROFESSIONS

Professionalism and ethics are difficult to define. In one dictionary ethics is described as the 'science of morals'. Certainly a code of ethics is a fundamental requirement for any profession, although remember that we are all prepared to rewrite our codes from time to time, depending on various circumstances, which are described and discussed at some length in Chapter 16. This chapter is taken from a paper by Lewis Anderson of the University of Greenwich in which he quotes Behrman when he says that Society needs professions more than at any time, even though it is in the process of destroying them.

Should we take this a step further in relation to the construction and property industries and ask the question, 'Is society prepared to **pay** for professional standards?' The question should not really have to be asked, since society cannot function effectively without professionals and the high ethical standards which they should bring to business life.

RELATIONSHIPS BETWEEN SOCIETY AND PROFESSIONS

Recently, the Association of Consulting Engineers carried out an historic review of the relationship between society and professionalism (drawing strongly on research work carried out at Bristol University). I would like to quote from it as it is applicable to construction professionals in general.

It says that a profession is generally recognized as claiming expertise as an over-riding need of society. For example lawyers, doctors and the clergy have looked after the basic needs of society from medieval times, followed by architects and at the time of the industrial revolution by engineers, and latterly by surveyors. Accountants also appeared at that time to look after industry's finances.

From the middle of the 20th century there has been a hyperbolic growth in new 'professions' as the result of a succession of revolutions caused by the introduction of the welfare state (social workers), the creation of enterprise culture (management consultants) and, most recent of all, the arrival of information technology ('knowledge' experts).

This evolution has been matched with changes in society's concept of what constitutes a 'professional', particularly so in the last four decades of the 20th century.

In the period 1900–1920, during which the profession of consulting engineering was conceived and organized, the concept was based on the classical relationship of trust between the professional and client. The concept was that the self-interest of individual professionals should always be subordinated to the client, society and the profession in general. Since that time, the Association's code of professional conduct has emphasized the need to serve the client and by implication society at large; it has also required responsible behaviour towards other professionals.

The professional is expected to hold some special knowledge and expertise required by the client. A report produced in 1970 by the Monopolies and Mergers Commission singled out the following common ethical characteristics of a professional:

- the ability to maintain impartial judgement and integrity in the relationship with a client;
- the ability to create a personal relationship with a client based on confidence, faith and trust;
- the avoidance of 'certain manners' of attracting business;
- the holding with others of a common sense of responsibility for the profession; and
- the collective self-regulation of standards of competence.

The setting of visible standards of conduct and ensuring adherence to them therefore supposedly divides the professional from the commercial supplier of goods and services. However, a major problem for those outside the profession is to understand what the ethics are and to be convinced that members of a profession behave in accordance with them. This has become increasingly difficult in the past two decades owing to the greater commercialization of the provision of consulting services and the entry into these fields of less-reputable practitioners.

In the period of rapid growth in 'knowledge workers' from 1970 to the present date, the above ideals have slipped quite alarmingly and the professional indulging in the business of the provision of financial services have largely through greed, competition and deregulation reduced society's concept of a professional. There are many examples where society has been ill-served by these new professionals, such as in personal pension planning in the period 1990–1993. As a result, the external authority formerly enjoyed by the traditional professions has become lost, and the professional is being made to account for their actions rather than having them taken on trust as in previous years. The modern clients are, in many cases, more knowledgeable than their predecessors and therefore feel able to challenge the performance of the professional adviser, possibly due to the performance criteria of the professional advisers being much more objective and set by corporate methods rather than by personal relationships.

The concept of self-regulation required a guarantee of acceptable levels of competence and conduct. The status enjoyed by professionals through the chartered status of their professional institutions led to some social esteem and, through the control of supply and the facility to set fee scales, a reasonably secure business style.

Rising standards of education within society have also challenged the justification of these apparent privileges and caused the Conservative-led government from 1979 onwards to deal with restrictive practices through deregulation and competition. The theory was that restriction of competitive behaviour led to the protection of the less competent with a consequential loss of high standards. Unfortunately, removal of this protection – when combined with competitive pressure in the open market – produced an inevitable erosion of standards of social responsibility.

Finally, the review states that in today's new society there is an urgent need to find a proper balance between ridding the professions of inefficient protectionism and maintaining standards of professional responsibility which society deserves and can (and must) afford.

Again, I repeat my previous statement that society needs professionals and high ethical standards.

In the following chapters, Dr Chris Hoogsteden from the University of Otago in New Zealand leads us into the issues of bribery and corruption. He quotes the case of the professional working outside his own country who succumbed to bribery. I wonder how many of us would have done otherwise given the same circumstances, or how many would consider that situation one of bribery, given the context of culture discussed in Chapter 16.

Chapter 19 is developed from research by Professor Roy Morledge and Keith Hogg at the Nottingham Trent University and summarizes undergraduates' perceptions of professional ethics. The results certainly make very worrying reading to anybody who has grown up with the concept of professional ethics being paramount. It says that some 15% of undergraduates (i.e. the new entrants to the profession) would be prepared to accept cash from a contractor and some 30/35% a paid holiday. It is worthy of note that 76% attribute their standards of right and wrong to the teaching and example of their parents. The Universities and professions have a strong part to play in changing these attitudes.

Some of the questions which you may like to consider while reading Part Four are.

1. Should we continue to maintain high ethical standards?
2. Does society want professionals?
3. Does society need professionals?

QUALITY

In the second section of this Part we have two chapters on Quality Management Systems. As mentioned earlier, quality is closely linked with professional ethics and fee competition. At an international conference of Consulting Engineers in 1994 in Australia it became clear that apart from the UK, very few consulting engineering firms have a fully certificated Quality Management scheme. In fact, the conference was told that there was only one firm in the whole of North America! Why is this? Why have so many firms in this country adopted BS 5750/ ISO 9000 Quality Management? Well of course the answer is easy; our government made it clear this would become a necessary requirement for any firm seeking appointment for government projects. For good or bad we now have Quality Assurance or Quality Management.

As we are all aware, Quality Management is designed to ensure that corporate systems are formally designed as well as being properly applied, but unfortunately have little to do with the technical quality or correctness of the work of the organization. If a firm is registered under a Quality Management system, then is it inevitable that they will subsequently make a move to Total Quality Management?

I suggest that we all have a fundamental problem to address. Do we offer the same quality of work for all our commissions or should we be prepared to provide various levels of quality? Should we be prepared to offer a higher level of quality to those who want it (and are prepared to pay for it) and the cheap and cheerful (but of course sound and safe) to those who want a low price. There is a world of difference between providing a professional service and buying a manufactured product ranging from a Rolls-Royce to a Reliant Robin, but the concept is the same.

It is very difficult to offer different standards from the same organization which is not based on a production line but on the relationship of its staff to its clients, but I suggest that this is the natural consequence of the move to fee competition and quality assurance. Fee competition assumes that firms can put a price on different individual parts of the service, rather than seeing them as a coherent whole, in that case it may be that practices would have to set up separate firms each offering a different service, with the obvious implications that would entail. This is an issue that must be addressed in the near future by professional service organizations and their clients.

BALANCING QUALITY AND PRICE

As mentioned, this links us into fee competition and here I would like to refer to Sir Michael Latham's Working Group in CIB which has developed a construction industry standard for assessing bids on the basis of price and quality. The system allows quality to be measured and balanced against price in the full assessment of the value of a bid. It is the responsibility of the client to decide their own weightings, for example, they may give a 20% weighting to price and 80% to quality for an innovative or complicated commission or, vice versa, 80% to price and 20% to quality for a routine task. At least there is a fair assessment of all that is offered by the consulting firm.

We have all heard of bids which vary by a factor of 3 or 4 for the same work from similar firms. Clients who accept the lowest price in this situation are deluding themselves, as they cannot get either the same quality of service, or the same amount of work, from the lowest bidder as from the highest.

The way forward must be to establish a proper brief for the work and then to invite offers from professionals bound by the highest ethical standards which can be assessed on the basis of quality **and** price.

As a final note in this introduction, we cannot undervalue the work of Sir Michael Latham in his review of the workings of the construction industry. If his proposals can be introduced, I believe we shall see a new climate in which quality, professionalism and professional ethics can once again play their part in the development of the construction industry of the future.

CHAPTER 16

The development of ethical standards

16.1 INTRODUCTION

This chapter looks first at the meaning of the term profession, and at the notion of being professional. The relationship between trades and professions, the importance of learning, knowledge and expertise, and the philosophy of commitment are considered in some detail. It then examines codes for professional conduct and protective structures, and explores ideas of competence, integrity, morality and business, paying special attention to the frequent dichotomy between personal values and business ethics. The fixed nature of moral belief is considered, alongside the notion of transformable morals, in the form of situation ethics. It concludes with a discussion of impartiality and the idea that vested interest is actually constructive.

We could choose to look at the topic of professional ethics in a highly practical way, identifying ethical conflicts, such as insider trading or health and safety related to accidents at work, and deal with them on a problem/solution basis. This would be of only limited use – it would provide no context, and no understanding of the significance of the wider issues. Conversely, we could look at ethical theory, naturalistic cognitivism, normative ethical theory, deontological theory and so on. This would be very unsatisfactory, leaving too big a gulf between essential understanding of moral concepts and practical contemporary applications. We will therefore use what I term the 'Lighthouse Approach'. We professionals in the construction and property industries can be considered, metaphorically, as ships at sea. The sea is our professional and commercial world. As ships, we pass to and fro, occasionally interact or respond to one another, and occasionally confront or collide. Each ship has its own system of navigation, perhaps using the stars, charts or lighthouses. The lighthouses are predictable beacons. They do not force the ships to obey, or to steer a particular course, but they enable the ships' captains to make informed decisions. After going, perhaps, too far in one direction, they enable ships to return to a more familiar position. The lighthouse is, in this metaphor, ethical theory. By knowing what is moral, or acceptable, we are more able to move freely without getting lost. We may be safe beyond sight of the lighthouse, but we will certainly be more vulnerable.

Four complementary extracts covering, respectively, meaning, reaction, complexity and understanding, and society and professions, set out the essence of any meaningful discussion on the topic of professional ethics.

Heller defined ethics:

> Ethics is the condition of the world. Chemical substances or organisms can exist without ethics, but there is no *world* without ethics. 'Worlds' is not the sum total of lifeless and living things but the *meaning* of all those things . . .

and Vardy states:

> There is a need for each of us to consider where we stand on moral issues that arise in the real world and to determine how we will react to the dilemmas they pose or, alternatively, to abandon any moral sense and simply to do whatever the law will allow us to get away with.

McNaughton adds:

> The demand for moral experts is growing. They are asked to sit on government commissions of enquiry, and on ethics committees set-up by hospitals and the like. There is also a feeling that each professional person should become his own amateur moral theorist; no training course in a profession is complete without a course on practical ethics and moral theory. . . . An introduction to the main problems of moral theory will serve both to reveal the complexities of many moral issues and to help people to understand their own thinking more clearly.

Durkheim completes our definitions:

> Since . . . society as a whole feels no concern in professional ethics, it is imperative that there are special groups in the society, within which these morals may be evolved, and whose business it is to see they be observed. Such groups are and can only be formed by bringing together individuals of the same profession or professional groups.

Heller is indicating that ethics are entirely philosophical and are not tangible rules. They are beliefs and judgements based on understanding and reason. This is a useful beginning, because too often a 'flow chart' approach to ethics is attempted. Our response to the philosophy of ethics is considered by Vardy. He differentiates between morality and law, and places firmly on the individual the responsibility for moral and immoral action. McNaughton develops these ideas and indicates how the complexity of the subject is not a disadvantage, but is a good basis for developing skills of understanding and thinking in professional life.

The necessity for, and protection offered by, groupings of like-minded individuals is explained by Durkheim. From Heller's philosophy through Vardy's and McNaughton's application and analysis, Durkheim offers the growth of professions and institutions as a logical and necessary response.

16.2 WHAT IS A PROFESSION?

It is likely that any author writing with the word 'professional' in the title will at some stage address the thorny problem of defining the term. It is equally likely that no definitions which are offered will be received with universal acclaim. Some authors attempt to define the nature of the term profession, not by description but by listing those occupations which they feel represent the meaning of the word. Bennion, while compiling such a list himself, acknowledges that it is a rash exercise. The Monopolies Commission in 1970 felt no such reserve, combining extensive lists with exhaustive definitions.

One of the most prevalent definitions of profession is 'the absence of trade'. The distinction between profession and trade is now much diminished in importance and is certainly obscure. Estate agency is a good example of the present interweaving of a profession with a trade. By any definition, much estate agency work is clearly that of a professional. The brokerage aspect of the work is clearly trade.

The growth of Market Panels and Skills Panels within the Royal Institution of Chartered Surveyors gives clear evidence of a pragmatic realization by the RICS that new working arrangements are needed to make the institution better able to support its members in an increasingly competitive marketplace. If there are no true professions now, Behrman asks, why should we concern ourselves with the issue? The answer is that society needs professions, even though it is in the process of destroying them out of a misguided ideological attachment to free enterprise and market-decisions.

16.3 WHAT MAKES A PROFESSIONAL?

The concept of a professional person is continually evolving. In several obvious ways, the construction industry is no different from many other walks of life in its use of the term professional. A good example, is the professional footballer. Footballers are regularly described by their managers as 'true professionals' when they continuously exhibit commitment to their club, usually off the field as well as on it, and when they set a good example to others. To merit the praise, they would also have to show their commitment in training and they would have to be the sort of person that takes some responsibility for their own achievements, as compared with another player who perhaps always has to be led or encouraged.

The same applies to any member of the construction team. To be classed as professional, they would have to show commitment, perhaps to the firm but more certainly to the project with which they were associated, and would be required to accept some degree of responsibility. The work undertaken should not be mechanical, and should require some degree of learning. A professional is usually held to have some special knowledge. However, it is in no way adequate to describe a member of the construction team as a professional merely because of the showing of some commitment to the task or assignment and having some special knowledge as a result of an ill-defined period of learning. That would cast the net over far too wide a range of people. Almost everyone on a construction site could, at some time during some day, on this definition, be classed as professional. Certainly the tea-lady would qualify instantly. She can be held to show commitment, and she holds special knowledge; that is to say, she alone knows rotas and teas-lists, and, assuming she was not born with this knowledge, she must have acquired it over some period of learning, even if it only took a day or so. It could be contended either way whether or not the task of a tea-lady was mechanical or otherwise. There is certainly a mechanical element in it, but then few tasks are completely cerebral. Even the architect has to draw countless standard details, and put standard annotation on every drawing.

If a person is required to be fully committed to a commission, and is required to bring to bear extensive special knowledge acquired over an extended period of time, much of it probably as a result of structured tuition and learning, and much of it primarily of a cerebral rather than of a mechanical nature, then such a person could claim to be considered a 'professional'. All of the professional institutions associated with the construction and property industries now have a well-documented commitment to continuing professional development (CPD), as recognition of the importance of continual learning.

Perhaps the key quality is commitment. Commitment is defined in the *Oxford English Dictionary* as 'an obligation undertaken; declared attachment to a doctrine or cause'. There are two important implications in this definition. Firstly, there is the sense of obligation, and secondly there is a declaration.

For a person to feel obligated, they must regard the execution of a task and the discharge of a duty as inextricably linked. Duty and task are not the same. The discharge of a duty is more fundamental that the execution of a task. A task can be executed without fully discharging a duty, and a duty can, in other circumstances, be fully discharged without the task being executed to completion. If a client commissions a surveyor, the surveyor has a duty to provide the client with a level of competent service at least equal to, and preferably exceeding, the client's justifiable expectations. The client's satisfaction does not necessarily determine the extent of the surveyor's execution of duty. Duty is absolute, but satisfaction is relative.

The declaration is of equal importance, but is more difficult to define. A professional can be held to have declared an attachment to a doctrine or cause. The doctrine or cause must either be specific to the profession (as with the Hippocratic Oath for doctors) or must be a general affirmation of fundamental principles universally applicable to a greater or lesser degree, across the whole broad spectrum of professionalism. A fruitful area for further consideration of these principles is fees, and the whole area of fee competition, which did not occur before fee scales ceased to be mandatory. The general declaration is an affirmation of integrity. A professional person holds themselves to be technically competent, committed, learned, and a person of integrity. They do not need to reiterate this, it is implicit. A professional person without integrity, in the sense of 'uprightness and purity' is an oxymoron. It would be highly imprudent to believe, however, that this contradiction does not occasionally exist. This leads to codification of integrity, in the form of Codes for Professional Conduct. For Chartered Surveyors, the backbone of the Rules of Conduct is Bye-Law 24.

16.4 CODES FOR PROFESSIONAL CONDUCT

Increasingly, the law regulates all aspects of business life. Whereas unwritten moral codes used to guide the actions of managers and professional men and women, today ethical codes, guidelines and the rule of law seem to be taking over. The danger of this, however, is that what is moral becomes what is legal. There would be little benefit in ascribing the label professional if so doing did not serve to enlighten the public as to the qualities to be expected. It acts on the one hand to advertise, to the public at large, the status of the individual, and on the other hand it acts as a reassurance to the public at large. The professional is relieved, to some extent, of having to convince individual clients, on a regular basis, of their integrity and trustworthiness.

Of course, the public must be reassured. Trivial comforting is not adequate. The professional person needs to have their status substantiated by credentials, and these take several forms. All professional credentials have one reference point, and that is exclusivity, resulting from some form of selection procedure. For registration, or membership of a professional body to be meaningful, registration or membership must be restricted. These restrictions need to relate to public perceptions of standards and also to professional aspirations, and take the form of codes.

The codes have three main functions. One is to provide present and future credibility for the organization. They afford values to the organization which otherwise would be lacking. This cannot be understated but it is intangible, and is a product of the other two functions. These other functions are to produce a yardstick standard for the organization a level above which many, it is hoped, will attain but below which none will be included, and also to provide a mechanism and structure within the organization for preserving these standards and for removing or censuring those members who fail to comply.

The public knows that the RICS, RIBA, ICE, etc. are professional and august bodies, whose members have attained a high standard and have been admitted to membership on a selective qualitative basis. It further knows that the members of these institutions are subject to codes of practice and professional indemnity insurance which are designed very much with the interests of their clients and the public in mind, and that if they fail to comply with these codes then they will be subject to sanction of some form.

> Membership of the (RICS) should be a big plus point. It imposes a code of conduct that has to be observed and obliges members, through CPD, to be informed of any legal changes and to generally keep in touch with movements in the property market. (CSM, 1995a)

As membership is discretionary, and sanction may bar a member from continuing as such, the code is enforced more as a preventive measure than as a retrospective judgement. This, too, benefits the public. It is little comfort to a client to learn that their errant surveyor has been struck-off after the client has suffered the damage. The prestige of membership acts as a positive incentive for the surveyor to maintain eligibility for membership of the Institution.

The previously mentioned yardstick standards take various forms, but generally include the provision of some method of qualitative assessment such as examination and assessment of professional competence. This is linked to a framework of structured learning, usually through higher education or CPD. The public is aware of these yardsticks even if it is ignorant of the precise mechanisms.

Codes, therefore, protect the public and the membership and also maintain credibility. The more august the body, the more keen it will be to protect 'the body' rather than the individual errant member. This is a very difficult area. Which has more credibility – the institution which never has the need to withdraw the privileges of a member, or the institution which boldly purges itself of undesirable members? In general terms the latter is probably the more credible, assuming that both the purged and the unpurged institution have exactly the same proportion of errant members within their ranks. Regular purging could very easily be misunderstood. It could look to an outsider that the one institution had a good membership, whereas the other had a regular undesirable element. It is no comfort to a client to know that 50 or 100 surveyors will be removed from the institution this year, in order to maintain standards. The client will understandably be wary that the 50 or 100 were members in the first place.

16.5 COMPETENCE, INCOMPETENCE AND NEGLIGENCE

There is a general consensus that all professionals should strive to make being competent the lowest threshold level, and should strive to ensure that they practise generally at a level above mere basic competence. Professional competence comprises **efficiency** (in doing a task economically), **sufficiency** (in providing a full service to a client) and **capacity** (which is the ability or capability to undertake the commission).

Incompetence is not an attractive criticism in professional circles. The mere suggestion or implication of incompetence can have an immediate and far-reaching adverse effect on the professional activities of the person accused. It is rare for a professional to describe another professional in such terms (in itself an unprofessional act), and a far more cautious and less damaging euphemism of negligence is usually employed. The accusation of incompetence hits at the very core of professional standing. A reputation carefully established over many years can be dismantled very rapidly when the competence of the professional is impugned.

Negligence and incompetence are often created as interchangeable, but this is not the case. Negligence is the want of proper care, and also the omission of such duty of care for the interests of others as the law may require. Negligence is a much more invidious accusation that of incompetence. To be negligent, the professional must have had the ability to be competent, but has disregarded the crucial importance of exercising this ability. The negligent professional wantonly disregards the course of action which is in the client's best interests. It may be wanton disregard for a number of reasons – for example haste, financial advantage, time or effort. Having disregarded the best interests of the client, and having encountered difficulties, the professional is clearly then open to a charge of negligence.

16.6 MORALITY, ETHICS AND BUSINESS

There are some professions where business life and private life are inseparable (for example, religious ministry). This does not apply equally to all professions. There is, for example, no unified public perception of how a registered architect should conduct themselves at leisure time. Certainly, some leisure activities may reflect unfavourably on their professional designations, and some activities may attract sanctions from the professional body. These areas are well defined, and there would be few occasions when a professional could unwittingly transgress them.

What is it that enables someone to adopt one persona in business and another in private life and yet enables that same individual to be respected in both areas? The answer is at the same time very simple and very complex. On the simple level, the public accepts that there is an element of any professional work which is technical, and subject to expertise. It also accepts that socially challenged individuals can be 'very good at their jobs'. There is an acceptance of this universally, and in extreme cases more credibility is given to a professional who is not socially accepted, than one who is. An example might be a surveyor who is flamboyant in the extreme, and in terms of normally accepted standards of professional behaviour, sadly lacking. Allowances may be made for such a character to the extent that their work is admired all the more because it is a product from such an unlikely source. Their virtues or abilities are extolled with the preface, 'You would never think it to look at them, but . . .'

The complex level is not as clear, for it is at this level that the ethos of professionalism is the dominant factor. A professional is always a human being – no machine anywhere today is likely to be ascribed the title 'professional', nor is it likely that any future machine would attract the description. Because all professionals are human, all the time, and professional for varying periods, the human factor is important. When criticising someone's professional endeavours, it is easy to forget that they undertook the commission, not as a machine, but as a human being.

As ethics and morality are products which involve aspects of right and wrong, it is implicit that ethics and morals may mean one thing to society as a whole, but may mean several other things to the individuals who make up that society. There is no need for deep philosophical thought here. It is, quite simply, a question of understanding. We all understand what makes an action right or wrong, and never so clearly as when the involved parties do not directly include us. When we are involved, we invariably temper our judgement of right and wrong according to our own perception of the facts, as we see them. We also make judgements based on the standards which we have either adopted or have been taught. These standards all relate to the common perception of morality and ethical behaviour, but they are not uniform. This is not for experts, but for us all, as ethics is too important to be left to the experts.

It is when business is undertaken that these subtle differences in standards have implications. Beauchamp and Bowie are very clear about these implications:

> It is common to hear people say that ethics is of little concern to business persons. It is said that ethics is ethics and business is business. The implications is that ethical issues do not matter in business. Such a view is so fundamentally mistaken that one might wonder why it is widely held. First, the practice of business depends for its very existence on the moral behaviour of the vast majority of citizens. Imagine trying to practice business in a society where lying, stealing, and other immoral actions were permitted. Business could not be practised in such a society, for, at a minimum, business requires a society where contracts are honoured and where private property is respected. Bribery, kickbacks, fraud, and monopolistic activities in the restraint of trade have all been judged inappropriate, because they are immoral practices. Second, there are and always have been special moral norms for business activity itself. These norms are usually devised by business persons. There are, for example, standards of good business practice. Sometimes these standards are written into special business codes of ethics.

It is quite usual to hear people suggesting that in fact there are two, equally valid, sets of moral codes. One covers our business activities and the other relates to our non-business or social activities. In business, people frequently make moral judgements, or carry out tasks based on ethical values, which differ substantially from the judgements or action they would make if the situation arose in private life. For example, a quantity surveyor may argue, quite forcefully, to reduce a rate for a valuation of a variation, when they are aware that they could argue as forcefully on its behalf as the contractor is doing. This same scenario is enacted frequently when negotiating claims. Few contractors would have the nerve or the resolve to submit a substantial claim for the net amount owing to them, in the belief or hope that this would be agreed in full by the other party. Inevitably, the contractor will submit a claim enhanced by at least the margin which they believes will be negotiated down by the employer's representative. This system of offer and counter-offer applies in many fields of life, and is subject to certain rules. It presupposes, for example, that both parties understand the fair value of the item being negotiated. It also requires that both parties are reasonably equally balanced, such that the fair valuation is the product of agreement and not a result of one party dominating the other.

The principles of the transaction, and negotiation, described above relate to all areas of business life, and are particularly prevalent in the construction industry. They are accepted business dealings. The acumen normally required to be successful in business is frequently held to be a skill separate from private dealings. In some cases this may be appropriate or at the very least acceptable, but that is not always the case. Morality and ethics are inseparable from a perception of right and wrong. If it is right to act with honesty, and wrong to deceive, then the difficulty of negotiating for the best advantage becomes evident. If a contractor knows that they are unlikely to receive a fair return without first overvaluing the work are they morally and ethically wrong to do so? If the employer's representative knows that the contractor will inevitably have overvalued the work, are they morally and ethically wrong to undervalue it, knowing that the eventual agreement will likely be fair? Do two wrongs, in fact, ever make a right? This is not a superficial moral tussle, for although the issues given in this example are simplistic, they can quite readily be developed into very complex issues. Custom in some parts of the world requires 'facilitation fees' which in other parts of the world are seen quite unequivocally as bribes, which is further discussed as a case study in Chapter 18. Should these 'fees' be paid? If they are not paid then no commission will result and therefore associated moral and ethical issues result not least in the redundancy of the unoccupied workforce. Is it an adequate ethical or moral response to accept such practices as convention, and 'only a business matter anyway' while paying lip service to ethical and moral values in private dealings? The public perception of a professional person is that they do not have dual standards, and do not wear morality like an overcoat, to be discarded as circumstances dictate.

16.7 INTEGRITY, DISCRETION AND RESPONSIBILITY

It has been shown that one of the most important qualities required in a professional is integrity – a basic inner honesty and uprightness. The pressures of the business world dictate that modification of these high principles and standards will occur, according to the situation. The extent to which these standards are modified depends primarily on three factors. First, if the pressures are very great, then the incentive to yield will be greater. Second, if the standards are not deeply rooted, then their inadequacy will be sought out by external influences. Third, the standards may be modified in spite of the fact that they are neither exposed to intense pressure nor only superficially held. They may have been modified under the discretion of the parties involved.

Discretion is a valuable aspect of a professional. Exercise of this discretion is one example of the expertise which sets a professional apart. Machines can perform tasks, but judgement is a human response. It is important, before discretion can be exercised, for the correct procedures first to be identified. These correct procedures are the traditional or routine elements that normally require little thought or modification. They are relevant to a wide variety of circumstances. Few professionals would need to deliberate over their implementation. Discretion therefore has little part to play in the general implementation of the quality of integrity. In its simplest terms, it amounts to a statement such as 'I am an honest person, and a professional of integrity. I know you have undervalued your work, and therefore I cannot accept your valuation but I must pay you the higher sum to which I know you are entitled'. Exercise of discretion in this way would be admirable, but could leave the professional open to criticism from a client who sees only that they are paying more than they may otherwise get away with, which would be an understandable reaction from a client.

What the professional has done, though, is to manifest these qualities which the client demands that they should have anyway, i.e. honesty and uprightness.

There are also other circumstances where the integrity of a professional may be best preserved by the introduction of discretion. **Situation ethics** are ethics based on the proposition that no conduct is good or bad in itself and that one must determine what is right or wrong in each situation as it arises. There are no absolute rules, as not only does the professional have much more opportunity (and indeed necessity) to exercise judgement, but is also more vulnerable in doing so. If there is a code or a rule, then a person can use that rule as a defence against criticism – 'I could not do such and such because there is a rule which forbids it'. The discretion of any professional at any time does not extend, for example, to acting outside the Common Law. Goldman states that:

> Legal limits are generally recognized. The question is whether businessmen ought to recognise moral obligations beyond requirements of law, when assumption of such obligations is compatible with maximisation of profit.

Remove these rules or codes and the professional can then be criticised either for exercising discretion, or for not exercising discretion. The extent of the criticism will probably be directly related to the success or otherwise of the decision made. The difficulty from the professional's viewpoint is that the decision will have been made on the basis of available facts, whereas the criticism will have the benefit of hindsight.

The area of situation ethics is a very difficult one. Many of the ethical values previously held need to be reassessed, for the rules change. Ethics, after all, addresses questions about moral choices, and exceptions to moral rules, and about the extent of moral responsibility. A clear example of this can occur with 'facilitation fees' referred to earlier. These are informal (though frequently structured) taxes applied by various involved parties in return for that party's cooperation in helping the contract to be placed. Sometimes these fees are secretive, but on some occasions they are remarkably overt. If a system of 'facilitation fees' is considered in a UK context, then the fees are seen, quite clearly and unequivocally, as bribes. Does this mean that the facilitation fee is always, in fact, just a bribe by another name, or does it mean that it is an acceptable remuneration in one society and a bribe in another? It is inevitable that conflict and opinion and accusations, will result from any discussion of this subject, just as conflict of conscience will also occur. Certainly, application of pure ethical values precludes participation in any oiling-of-the-wheels to obtain work. If it is wrong to bribe and to induce, then pure ethics dictates that it is always wrong to do so. There can be no exception. Where the custom in another society varies from the home custom, pure ethics will need to be applied either to the home custom or the custom of the other society. Spaemann supports this notion when he says

> The fact that ethical codes are to a large extent culture-relative is very often cited as a reason for rejecting the possibility of moral philosophy, that is to say, the possibility of reasoned discussion about the meaning of the word 'good' in an absolute rather than a relative sense. But what this sort of argument fails to take into account is that moral philosophy does not have to disregard this fact in order to function.

There is no choice. Ethics are a personal statement of morals, a set of behaviour rules related primarily to the environment in which they have evolved. The professional cannot exhibit and cultivate the principles of pure ethics in one society, and act with different ethics in another society with which they are temporarily associated. Pure ethics do not travel very well.

Situation ethics appear superficially to be the answer, but there is a price to pay. The advantage of pure ethics from a professional person's point of view is that they are readily identifiable by them, and they add to the person's standing and say much about them. Situation ethics are similar, but have a fundamental difference. The professional who is an exponent of situation ethics is less identifiably predictable. Let us imagine an overseas project, valued at several million pounds, which is half completed. A local official, with the power to disrupt, and perhaps even to cause the suspension of the project, demands a £50 facilitation fee to enable the contract to proceed. Pure ethics would demand that this bribe is denied, come what may. Situation ethics may well demand a more pragmatic approach. After all, £50 is a trivial sum of money to pay for ensuring millions of pounds of cashflow. It is, however, to some extent, a large moral price to pay. Few people would take anything other than a pragmatic approach to the problem illustrated. To resist on ethical grounds would cause distress to employees, to local people, to the local economy and probably prejudice the contract; £50 could be seen as a very small price to pay for alleviating such problems.

No sooner have we decided that it is obviously right to be pragmatic in this example than that opinion is challenged if we consider another circumstance in which the same official asks not for £50, but for a £50 000 facilitation fee. Is the situation the same – do the same ground rules still apply, and if not, why not? One reason is that situation ethics make proportion a large factor in any decision process. The responsibility for taking any ethical decisions is clearly apportioned, and the responsible person is also charged with maintaining the proportion of all the decisions made. The exercise of ethical discretion is never more important.

Williams does not allow the 'when in Rome' rule to be used as an excuse. He states, 'None of this is to deny the obvious facts that many have interfered with other societies when they should not have done; have interfered without understanding; . . . it cannot be a consequence of the nature of morality itself that no society ought ever to interfere with another, or that individuals from one society confronted with the practices of another ought, if rational, to react with acceptance'.

16.8 VESTED INTEREST AND IMPARTIALITY

The extent to which ethical discretion will be exercised will frequently be related to the nature of the vested interest of the party concerned. However impartial a surveyor holds themselves to be, or however much an architect or contractor tries to put their professional work above any self-interest, it is highly unlikely that vested interest can be vanquished completely. An architect may strive, almost self-effacingly, to satisfy the requirements of a client, but is never able to escape the reality that they will be judged professionally by the world at large on the basis of the finished development. They have a vested interest in ensuring that the finished development makes the sort of statement with which they wish to be associated. A finished building is in many ways the architect's business card. A contractor may strive very hard to satisfy the employer and to provide them with a capital asset with which they will be delighted. Never far from even this contractor's mind is the thought that there is a 'bottom line' to such an attitude. Goodwill in many circumstances has a calculable monetary value. Goodwill extending beyond this 'bottom line' affects directly and adversely the contractor's profitability. They obviously have a vested interest in the profitability of the organization and the satisfaction of their shareholders, if any.

Vested interests are those interests which are already established, and are therefore inherent within an organization or a project. It is very unlikely that any circumstances ever occur where vested interests are absent. It is the suppression or control of these vested interests which results in the varying degree of impartiality which can prevail. Impartiality is not a suitable or desirable aspiration for all parties in a construction team. Impartiality and disinterestedness are close companions. It is frequently far more productive and beneficial if there is an overt element of vested interest exhibited in a project. It is good for a contractor to strive to be profitable – it engenders a sound commercial attitude which can be harnessed to advantage by a shrewd design team. It is good for an architect to try to preserve the integrity of a design, and while not enforcing it upon the client, certainly perhaps explaining and defending it more coercively than they would otherwise be likely to do.

There is no virtue in striving to be impartial for the sake of it. Impartiality should be tightly controlled and carefully exercised. The most important member of the construction team, the client, is the least likely ever to be impartial, and the one most likely to have obviously overt interests.

16.9 REFERENCES AND FURTHER READING

Batts, M. (1995) The RICS and consumer protection; in *The Ivanhoe Career Guide to Chartered Surveyors*, CMI London, pp. 9–10.

Beauchamp, T.L. and Bowie, N.E. (1979) *Ethical Theory and Business*, Prentice-Hall, New Jersey.

Behrman, J.M. (1988) *Essays on Ethics in Business and the Professions*, Prentice-Hall, New Jersey.

Bennion, F.A.R. (1969) *Professional Ethics: The Consultant Professions and Their Code*, Charles Knight, London.

CSM (1995a) Leader, February, p.16.

CSM (1995b) Article, February, p. 9.

Durkheim, E. (1957) *Professional Ethics and Civic Morals* (translation by Cornelia Brookfield), Routledge and Kegan Paul, London.

Edge, J. (1991) Quality Assurance: CPD and the RICS, in *The Ivanhoe Guide to Chartered Surveyors* (ed. L. Parkin), RICS, London, pp. 85–90.

Glenn, J.R. Jr (1986) *Ethics in Decision Making*, John Wiley & Sons, New York.

Goldman, A.H. (1980) *The Moral Foundations of Professional Ethics*, Rowman and Littlefield, New Jersey.

Green, R.M. (1984) *The Ethical Manager: A New Method for Business Ethics*, Macmillan, New York.

Heller, A. (1988) *General Ethics*, Blackwell, Oxford.

Holmes, A.F. (1984) *Ethics: Approaching Moral Decisions*, InterVarsity Press, Illinois.

McNaughton, D. (1988) *Moral Vision: An Introduction to Ethics*, Blackwell, Oxford.

Monopolies Commission (1970) *Professional Services Part II: The Appendices*, HMSO, London.
Pattison, M. (1995) History and growth of the profession, in *The Ivanhoe Career Guide to Chartered Surveyors* (ed. L. Parkin), CMI, London, pp. 2–5.
Spaemann, R. (1989) *Basic Moral Concepts* (translation by T.J. Armstrong), Routledge, London.
Vardy, P. (1989) *Business Morality*, Marshall Pickering, London.
Williams, B. (1972) *Morality: An Introduction to Ethics*, Cambridge University Press, Cambridge.
Willis, C.J. and Ashworth, A. (1987) *Practice and Procedure for the Quantity Surveyor*, Collins, London.

CHAPTER 17

Are ethical standards good for business?

17.1 INTRODUCTION

This chapter considers the current relevance of professional ethics to the management of professional practices. Distinguishing professional ethics from business ethics, it considers whether adherence to codes of professional ethics is appropriate for the long-term commercial success of a practice, or represents merely the acknowledgement of a general constraint exchanged for the mandate to practice.

17.2 BUSINESS AND PROFESSIONS

The relationship between business and Anglo-American professions has become increasingly close, especially with the development of codified expertise in business management. An American sociologist, C. Wright-Mills, neatly summed up the position of professionals in business over 40 years ago:

> United States society esteems the exercise of educated skill, and honours those who are professionally trained; it also esteems money as fact and symbol, and honours those who have a lot of it. . . . When we speak of the commercialization of the professions, or of the professionalisation of business, we point to the conflict or the merging of skill and money.

For many professional practitioners, the potential conflict between skill and money is experienced in the form of two imperatives. The commercial imperative, to maximize profits or to contribute to long-term profit maximization, may seem separate from any moral questions. The second imperative is to serve clients' interests, within the general public interest, by the use of professional skills and expertise.

These imperatives may be seen as compatible, or sometimes in conflict. For our purposes, we would distinguish between two extreme types of practitioner. We would term 'liberal' practitioners as those who see professional practice as a form of public service. They would oppose any use of their expertise against the public interest, even if they are personally disadvantaged. Such practitioners are perhaps common to the 'liberal' professions, which may be less influenced than others by commercial pressures. This apparent distance from commercial imperatives is reflected in Bennion's classic study of professional ethics, which will be closely considered later.

The second type of practitioner, which we would label 'robust', would consider the liberal practitioner to be unduly moved by sentiment. While robust professionals may not necessarily regard the single-minded pursuit of profit as the only virtue, they tend to see professional work as objective and technical, and so unaffected by questions of ethics. Thus:

> . . . the ideology of the profession is often converted into 'maxims of everyday life' that need not be examined, reassessed or criticised too frequently or too closely. (Burrage *et al.*, 1990)

We would suspect that this is a significant, if not dominant, view within construction and property, with its strongly technical focus. Ethics thus appear to be theoretical abstractions, irrelevant to practice management, with the espousal of ethics merely the rhetoric of professional bodies. This perspective finds an odd resonance in that of critical social scientists who consider ethics to be part of an ideological cloak for undue privilege.

Rather than to pursue a normative view favouring one type of practitioner's approach over the other, our attention is focused on the question whether successful practice still requires close adherence to professional ethics, or whether, as robust practitioners might suggest, ethics can hinder the delivery of an effective service to clients.

17.3 PROFESSIONAL ETHICS, CORPORATE CONDUCT AND BUSINESS ETHICS

The core of professionalism may be taken to be that defined by Johnson, i.e. the possession and autonomous control of a body of specialized knowledge, which, when combined with honorific status, confers power upon its holders. Ethical problems lie with the limits which should be placed on the exercise of that knowledge-based power by practitioners.

The robust practitioner would not tend to perceive problems with privileged knowledge, assuming that specialized skills and competencies are exercised in clearly understood circumstances. At its most extreme, any professional action can be justified so long as it does not transgress legal requirements; thus, it should not invoke criminal proceedings, should not enable actions in negligence to be brought, and should not offend the (quasi-legal) codes of conduct of professional bodies. Within the considerable room for manoeuvre set by these legal requirements, the key criterion is what is most expedient, usually the most profitable course for the practice.

Much more limited space for action is provided by the adherence to professional ethics espoused by the liberal practitioner. If practitioners refuse to act for clients whose instructions may be contrary to such ethics, some profitable work must be turned away, and a viable service can only be offered to a limited range of clients. Practical problems may arise with public perceptions of the professional use of privileged knowledge and understanding, whatever the extent of adherence to the substance of professional ethics. Practitioners could be perceived to use this privileged knowledge and understanding to put their own interests before those of a client, or one client's interests before those of others, or clients' interests against before those of the wider society. A robust practitioner might feel that public or client perception is the only issue, or that the choice of action is one for individual consciences rather than the general conduct of professional practitioners.

Some commercial justifications for professional ethics may be found in the use put to business ethics. The study of business ethics and corporate conduct has mushroomed since the early 1970s, partly prompted by commercial pressures which recognized that, according to Marshall:

> . . . the whole public standing of a business organisation could affect its economic well-being. If a firm had a good reputation for the overall quality of its products, for its sympathetic handling of relationships with customers, for being a good employer . . . then this standing would cause the general public to be more favourably disposed to buy the products of the firm.

Not only may the conduct of corporate bodies have significant commercial implications, so may the conduct of those employed by such organizations. Business ethics deals with problems created by the individual consciences of employees, which can vary greatly in impact. As corporate conduct imposes rules which may over-ride individual moral sensibilities, business ethics may be seen and experienced as an attempt to extend employer control over individual behaviour.

Some students of business ethics suggest that integrity is an essential component in successful business operations. However, the main concern, implicitly or overtly, is with bad publicity, with attention to ethics being prompted by the need to avoid scandals such as insider dealing.

Professional ethics do not correspond directly with business ethics, but there are many similarities. The over-riding ethics here are imposed by an external profession on employees, but also on employers, some of whom may resent the ability of external professional bodies to impose guidance on their employees. Alongside this attempt to control individual behaviour may be set attempts to influence the conduct of professional practices, which may not only seek profitable work but also work which does not contravene ethics. As with business ethics, obviation of the dangers posed by adverse publicity may be a central concern.

However, a business which makes products or services whose parameters and qualities can be closely specified and controlled is in a very different position from a professional practice, whose services involve uncertainty in specification and outcome. Large corporate businesses can usually formulate clear policies for the future work of the organization, and to develop personnel and control their conduct. However, smaller practices may not be sufficiently resourced to devote much effort to the formulation of an appropriate moral stance, and be less able to calculate with any precision what might be gained or lost by specific policies.

17.4 PROFESSIONAL ETHICS, PUBLIC RELATIONS AND PRACTICAL DILEMMAS

The ethics which each individual profession is meant to formulate and follow vary between professions, since, as Durkheim notes,

> '. . . there are as many forms of morals as there are different callings.' Thus only broad general principles can be applied to the professions in property and construction, which have each developed specific applications autonomously. This ambiguity provides much space for alternative interpretations of particular ethical imperatives.

One attempt to set out a general ethics of professionalism was Bennion's somewhat didactic and uncritical study. This was written before an era dominated by Wright-Mills' 'merging of skill and money', and while it may appear foreign to practitioners fighting for survival in a strongly competitive commercial environment, its general principles are instructive.

Some practitioners would consider the following of professional ethics to be implicit and unproblematic, simply because they may correspond to suitable public relations. Bennion's six abstract professional ethics can, in outline, each be regarded merely as components of a suitable professional image. Thus, an image of competence is clearly essential, while clients will shun a practitioner who does not maintain discretion over confidential information. The ethic of humanity implies a sympathetic interest in clients' problems, while the 'cab rank' rule provides that no prospective client will be turned away without good reason. The impression of impartiality, whereby there is no conflict between one client's interests, or the practitioner's, or any other clients' interests, is again important, as is the notion that practitioners will take full responsibility for their actions. Finally, many clients will be attracted to a practice which values integrity, treating clients fairly with fearless attention to their interests. These ethical imperatives both fit a robust view of common-sense practice, as well as providing promotional rhetoric.

An area of practice common to most construction and property work is the conducting of negotiations, whether between landlord and tenant, vendor and purchaser, or parties in the construction process. Parties in negotiation rarely share the same position of power, and the manipulation of their respective knowledge and ignorance is often germane to the process. Bennion finds it difficult to reconcile ethics and the conduct of negotiation, 'a difficult practice', in which

> An element of bluff is inseparable from negotiation, and therefore cannot be condemned. How far is it legitimate to go? . . . Perhaps the last word should rest with Lord Esher, who said that how far a solicitor might go on behalf of his client was a question far too difficult to be capable of abstract definition, but when concrete cases arose, everyone could see for himself whether what had been done was fair or not.

Lord Esher's judgement could hardly provide less clear guidance, and provides no guidance over the dilemmas involved in negotiation. Does the fairness which characterizes the ethic of integrity mean that the ignorance of the other party or their professional representatives can be exploited? To what extent can misleading representations and threats be made during negotiation, or unnecessary delay (if it benefits the client) created? The commercial approach in both cases will usually be to favour the client's interest above any notion of integrity. If a practice is negotiating with another practice which is better informed and more competent in that area, should the client be informed? In such a case, not only is the client liable to instruct a more competent (or less honest) practice, the possibility of future instructions must be severely prejudiced.

Negotiations over the value of property notoriously involve comparable evidence which is often obtained in confidence; sometimes disclosure will be legally prohibited by confidentiality clauses. Bennion states that no use should be made of such information, as it breaches the ethic of discretion, but to fail to use it may achieve a poorer price for the client, breaching the ethic of responsibility. Finally, negotiations with one party may involve conflicts of interest with other parties. For instance, if a practitioner was negotiating on behalf of a succession of clients with the same party (a large contractor, or a purchaser of many properties), it would be tempting to soften early negotiations so as not to prejudice future dealings. Here the ethic of impartiality would be breached, between the interests of different clients and between the interests of the practice and the clients.

If the ambiguity inherent in interpretations of professional ethics prevents the provision of clear guidance, the strictly commercial approaches of the robust practitioner use very different, yet clearer, criteria. When negotiating, one criterion for conduct could be whether further negotiations are anticipated with the same party, who may later be in a better negotiating position. Expediency may demand that an early settlement can be profitably reached, avoiding the costs involved in prevarication. A final consideration is whether the negotiations could somehow provide damaging adverse publicity. This is not then a matter of doing good, but creating goodwill, possibly only when it is important and profitable.

Ruthless practitioners might appear to achieve superior outcomes than those who insist on strict adherence to professional ethics. Bennion asserts that the ethic of integrity demands fairness not only to clients, but also to the others involved, citing the imperative for architects and quantity surveyors to act impartially towards contractors. It is possible, however, that the pursuit of fairness to others may betray the trust placed by a client that the practitioner will obtain the optimum outcome. It may be fairer always to deal with reputable contractors, but they may face the client with higher costs; while better results may be obtained from 'strong-arm' debt collectors than more polite means of debt recovery. Not only may the client lose out, there may be the suspicion that the Practice is looking for the easiest course of action for itself. Ambiguity lies here in the inability of the client to determine what the consequence of an alternative course of action would have been. Both trust and ambiguity are inherent in professional client relationships.

17.5 PROFESSIONAL ETHICS AND CORPORATE CLIENTS

Abstract discussions of professional ethics tend to assume that individual practitioners have full control over their work, and are capable of exploiting the limited knowledge and understanding individual clients whom they serve. Ethics thus seem to serve to protect consumers against producer interests. However, in reality many professionals in construction and property deal with corporate business clients, and only serve the general lay public in a broad abstract sense.

These corporate clients may seek to impose their own values upon practices which they choose to instruct. These may be more concerned with a practice's other involvements, rather than its stance on professional ethics. For instance, they may not wish to be associated with other clients for whom a practice acts, which might be connected (say) with oppressive regimes or a damaging environmental record. No absolute standards can be anticipated or specified, as the objections may depend on the image which the client is seeking to protect or project.

Clients may place pressure on practices on quite different grounds, not necessarily in liberal directions. The incidence of religious, political, club or society connections with firms and their employees is commonplace and long-standing in Britain. Similarly, while a practice which was reputed to endorse strong-arm debt collection or ruthless negotiations might deter some clients, others might be attracted by its ability to achieve results. The development of mega-lawyering in the US suggests that large corporate clients can pressurize whole sections of professions to abandon their independence and any duty of wider public service.

Whether this has yet been manifested in construction and property is a question for empirical investigation, but the rewards involved in bending acceptable practice (and perhaps the interpretation of professional ethics) to the will of large clients may be considerable. It is perhaps not too fanciful to imagine a bifurcation between professional practices willing to act for clients whose requirements could offend professional ethics, and those which would refuse to act. The latter would have to choose to service smaller and less remunerative clients, and although the maintenance of independence and probity might ensure their long-term survival, such practices might prove to be smaller and poorer in the short term.

17.6 CONCLUSION: GOOD FOR BUSINESS?

The uncertainties involved in their application suggest that general professional ethics provide no clear guidance for everyday practice. It may be that each individual practice, within the rules of their governing professional body, needs to determine their own rules, which in turn may assist their attempts to define the kind of client and instructions to be sought or rejected.

The specification of professional ethics may benefit a profession, but not individual practices. Those who abide by the rules may find themselves outflanked and undercut by those who do not. If ethical practice involves refusing to act, one consolation is that much of the offending potential instructions might prove troublesome to handle, and could damage the Practice in the long term. This may apply when adverse publicity may damage the reputation of a Practice, although much depends on whether misdemeanours will be discovered, and the

Practice's view of suitable clients. Their scruples and their ethics, rather than those of the profession, may dominate.

As such, there is no evidence that moral excellence necessarily corresponds with effective and profitable practice. A reputation for ethical practice, rather than for high levels of professional effectiveness, may deter more clients than it attracts. This is particularly the case with large clients which may seek close control over instructed practices. In such cases the Practice may have to take a strategic decision whether to seek or to avoid such clients.

While there are ambiguities in the application of professional ethics, commercial imperatives force any Practice to adopt screening and filtering processes over initial and continuing instructions. These may combine ethical criteria with those of commercial viability. Any Practice must determine whether it has (or could develop) the competence to deal with an instruction, and whether the work involved will prove profitable. This initial screening will filter out some instructions which would offend ethics, and for others, the additional risks involved, of adverse publicity, or refusal to pay fees, will provide a further screen which would be reflected in profitability. This leaves a minority of cases in which damaging publicity is unlikely to result, but ethical objections arise. Here it may not be 'good for business' to refuse to act, although it remains possible that such cases will be more difficult to handle than cases with unambiguous features.

While it is difficult to be prescriptive, it would be advisable for all Practices to develop procedures to deal with the ethical implications of instructions. Procedures will need to determine the ethical criteria upon which instructions would be refused or terminated, including those of business ethics, and their precise interpretation for the Practice. Everyday operational rules could be derived from these criteria, including an early warning system to alert the Practice when problems could arise.

Our stress on the ambiguity of interpretation of professional ethics may mean that absolute imperatives cannot be specified and followed, but this does not endorse a relativist approach to ethics. It is better to engage in discussions and debates over ethics, applied to practical dilemmas, even if opinions and perspectives differ. The alternative is to surrender such problems to commercial imperatives, even if the latter do filter out some ethical problems, and to caprice and chance.

17.7 REFERENCES AND FURTHER READING

Bennion, F. (1969) *Professional Ethics: The Consultant Professions and their Code*, Charles Knight & Co., London.

Bowie, N. (1991) Business ethics as an academic discipline, in *Business Ethics: The*

State of the Art (ed. R. Freeman), Oxford University Press, London, Chapter 1.

Burrage, M. and Torstendahl, R. (eds) (1990) *Professions in Theory and History*, Sage Publications, London.

Burrage, M., Jarausch, K. and Siegrist, H. (1990) An actor-based framework for the study of the professions, in *Professions in Theory and History* (eds M. Burrage and R. Torstendahl), Sage Publications, London.

Collins, R (1990) Market closure and the conflict theory of the professions, in *Professions in Theory and History* (eds M. Burrage and R. Torstendahl), Sage Publications, London.

DeGeorge, R. (1991) Will success spoil business ethics?, in *Business Ethics: The State of the Art* (ed. R. Freeman), Oxford University Press, London, Chapter 2.

Galanter, M. (1983) Mega-law and mega-lawyering in the contemporary United States, in *The Sociology of the Professions* (eds R. Dingwall and P. Lewis), Macmillan, Houndsmills.

Johnson, T. (1984) Professionalism: occupation or ideology?, in *Education for the Professions* (ed. S. Goodlad), Society for Research into Higher Education/ NFER-Nelson, Guildford.

Marshall, E. (1993) *Business and Society*, Routledge, London.

Walton, C. (ed.) (1977) *The Ethics of Corporate Conduct*, Prentice-Hall, New Jersey.

Wright-Mills, C. (1951) *White Collar: The American Middle Classes*, Oxford University Press, London.

CHAPTER 18

Case Study 1: the ethical experiences of surveyors in New Zealand

18.1 INTRODUCTION

> Professionalism is not an ethereal something understood only by philosophers and advocated only by idealists. It is the intangible, yet very practical cloak of integrity, altruism and culture that shrouds and protects a body of men [sic] whose conscious purpose in life is to serve humanity. (Robinson, quoted in Simpson and Sweeney, 1973)

> But, you see, I can believe a thing without understanding it. It's all a matter of training. (Lord Wimsey, in Sayers, 1931)

The tabloid exposures on unbecoming behaviour by professionals seldom involve land surveyors. After all, the various exploits of dishonest politicians, corporate high-flyers, and other occupations are invariably more newsworthy, for example, the well-publicised rise and demise of Messrs Nixon, Maxwell and Boesky. Perhaps, even, the vocational title itself, and the workplace activities of land surveying, are rather too prosaic for the investigative journalist anyway! Yet, as with most professional institutions, surveyors world-wide do place standards of ethics, and an adherence to a specified code of conduct, high on the list of expected professional practice and behaviour.

More recently, and for good reason, formal education in ethics has started to appear in various professional undergraduate curricula including those of medicine, engineering, accountancy and law. Closer to home, the author has had an intimate involvement with the design, provision and delivery of a substantial professional practice course for undergraduate surveying students over the past decade; a course which has specifically included aspects of ethical theory together with various case studies concerned with (un)ethical behaviour.

Arising from this educational involvement, and also a strong interest in professional issues generally, it became clear that there has been scant research into the real-life experiences and perceptions of professional surveyors on ethical matters beyond the occasional regulatory investigation into shoddy fieldwork or overcharging. An attempt to remedy this lack of knowledge led to an exploratory study of 600 New Zealand land surveying practitioners (190 responded) which was aimed at determining:

- which specific practices are viewed by practitioners as unethical;
- with what frequency do these practices occur;
- the extent to which land surveyors experience conflicts between their personal ethics and those of other individuals and/or the organizations where they work;
- the comparison of present ethical standards with those of the past; and
- the collective strategies concerning unethical behaviour which a professional organization might consider adopting in the interests of the public.

The responses to such questions from the 190 practitioners have been reported in academic journals, with some of the more interesting results being that:

- perceptions of what constitutes unethical conduct range widely with the main categories being those of unprofessional practice (e.g. knowingly undertaking work beyond one's level of expertise), unethical policies (e.g. obtaining unfair advantage by using insider trading) and unethical financial practice (e.g. collusive pricing);
- a significant proportion (63%) of respondents felt that the standard of ethical behaviour had declined since the time when they entered the profession;
- the same proportion (63%) predicted that ethical standards would decline further;
- the most common stated cause for this 'unhappy state of affairs' was that of a general decline in both community and commercial standards; and
- suggestions for improving professional standards ranged from the draconian (life suspensions and floggings were included!) to the thoughtful (better auditing, improved Continuing Professional Development and education in morals and ethics).

One area of the study which has not been reported on before is the pragmatic issue of real-life ethical dilemmas and how surveyors cope with these situations when they occur. Of equal importance, however, is the formal process by which surveying practitioners in New Zealand are actually informed about the specific type of behaviour which the collective body regards as inappropriate and would warrant investigation with possible punitive action. This paper considers both of these matters before closing with the responses of this same community to a hypothetical case of bribery.

18.2 PART 1: A FORMAL PROCESS

Existing models for ensuring ethical professional behaviour appear to operate on the assumption that once ethical codes and standards of conduct are decided upon, and put in place through published guidelines, life is reasonably straightforward. Some actions are then clearly good – others are obviously bad. Unfortunately, the real world is often a shade of grey and making rules, and monitoring professional behaviour to ensure adherence to such rules, can be a rather sticky and usually thankless task.

The most recent professional Code of Ethics adopted by the New Zealand Institute of Surveyors (NZIS) is quite succinct in exhorting its collective membership that:

> All members in pursuance of their livelihood and vocation are to conduct themselves having regard to high ideals of professionalism, courtesy, personal integrity and public service in a spirit of fidelity to clients, employers, and employees and of fairness to fellow members.

No definition is provided for any of the terms contained in this dictum and, at first glance, professional surveyors are simply left to interpret for themselves whether a contemplated action could or should be viewed as unethical, say on the grounds of courtesy, personal integrity or public service. However, some guidance for practical life, albeit a little circular, can be extracted from the Institute's stated Code of Professional Conduct (Rule 21) which requires that:

> No member shall act or take part in or associate with or allow their name to be associated with any enterprise or action derogatory to the profession of surveying or which may constitute unprofessional conduct.

Here, it is worth noting that the Council of the NZIS has in the past published a list of those matters which were considered to:

> . . . reflect appropriate professional conduct or which might constitute unprofessional conduct for the administration of this rule and Rule 20.1.

A more recent listing of those matters which are considered to be encompassed by the term *Professional Conduct* is provided as Appendix 18A (p. 257). As can be seen, and not unnaturally, the list includes various sensible topics and issues which practitioners should be aware of such as: confidentiality; the acceptance of favours; and, the recognition of personal frailties, e.g. technical inadequacy. There are also some quite fascinating 'motherhood' statements.

It is also of interest that although registration as a surveyor can be suspended (or cancelled) and a fine not exceeding NZ$ 2000 can be imposed for unprofessional conduct, this particular term is nowhere defined. Rather, the approach which is seemingly taken is that unprofessional conduct is simply the antithesis of professional conduct (as described in Appendix 18A; see p. 257). If this is the case, then there could yet be some interesting ramifications as miscreants are struck off for failing to '. . . exchange information and experience with other members' or when those who have neglected to '. . . contribute to professional publications' are castigated. (Of course, editors of professional journals who frequently struggle to extract papers from busy practitioners might be delighted at such a turn of events!)

Perhaps a rather more important point to be noted in regard to these Codes of Ethics and Standards, and one presumes here that they are not especially unusual in either tone or content, is the lack of clear practical statements. For example – and this particular theme will be returned to later – what should a member actually do if confronted by an issue of bribery when overseas? Taken together, the various codes and expectations discussed above are somewhat equivocal and, frankly, of no great assistance to individual practitioners in dealing with such matters. After all, given the requirements in the Code of Ethics as recorded above, what should young surveyors do when the interests of the client and the spirit of public service actually collide?

18.3 PART 2: ETHICAL BEHAVIOUR IN PRACTICE

For most professionals, the task of judging another's actions as unethical is difficult enough; admitting to situations of temptation towards, and sometimes the commitment of, unethical practices oneself is probably more so. Given all this, and with the clear promise of confidentiality, respondents to the survey were asked to share their own personal experiences in terms of ethical dilemmas by being asked:

> In the past, have you personally experienced serious conflicts between the expectations of you as an efficient, cost-conscious surveyor and your own professional ethics?

Some 38% of the respondents to the questionnaire (namely 72 surveyors) responded positively to this question. While this figure may seem rather large, it should be remembered that these surveyors were essentially being asked to look back over their entire careers. For some of them this may encompass several decades. Using a variant of the critical-incident approach, the subset who responded positively were then asked:

> Please could you briefly describe the general circumstances of the **most meaningful occasion** and how you resolved it.

Resulting from this question, the main areas in which these professional surveyors have had to deal with personal ethical conflicts have been those of economic issues (36%), surveyor–client and surveyor–surveyor relationships (31%), and technical matters (23%).

In order to provide further analysis of the responses from a different perspective, a schedule devised by White and Rhodeback was adapted for categorising ethical dilemmas in the following manner.

18.3.1 Value and goal conflict

This results from ambiguously defined goals and can be manifested when services or resources are withheld on either side.

Many of the described ethical dilemmas were the result of the perceived conflict between various values and goals, with both those held by the individual surveyor and, on other occasions, with those of other persons involved. For example, there are those cases where the surveyor has considered that there is a clear conflict of interest because of being asked to take on more than one role:

> Minor conflicts of interest occur mainly in the field of being asked to certify engineering works that have been designed by us, approved by Council but constructed without our knowledge or supervision.

In a similar instance provided by another respondent, the coping mechanism was quite straightforward and collegial:

> I referred work to our opposition firm when approached to certify our own setout as independently checked.

Similar feelings about the appropriate course of action were provided by another surveyor:

> As a local authority surveyor allowed to undertake a limited amount of private work, I was asked to undertake a survey for which another surveyor had prepared the scheme plan [*typically an application plan for subdivision*]. I had no hesitation in declining to undertake same.

However, this type of solidarity with colleagues was not always quite as forthcoming and several surveyors reported on situations similar to the following:

> While working overseas a client requested a development to be conducted in a manner detrimental to the environment. I refused to comply and as a result a competitor was employed. A further long term result was that my licence to practice in a particular area was withdrawn by the Surveyor-General in favour of my competitor. Moral is short-term gain, long-term pain.

For others, there has also been a feeling of being let down by the system.

> [There was] an occasion where the adopted line of Mean High Water Mark was quite definable and I knew that illegal reclamation had taken place not accidentally but quite deliberately. I refused to 'bend' the line to eliminate the problem. Another firm took over the job from me, without consultation and obliged. It passed the Council and was accepted by DoSLI. (DoSLI is the Department of Survey and Land Information, the regulatory agency)

The need to take a hard stand on principle, even where the required action may actually have been within the law (technically at least), was another situation which led to two surveyors declining to continue with the work and turning away clients forthwith.

> A client requested me to make a planning application for a subdivision based on a specific land use but also advised me that he had no intention of carrying on with that use if a consent was granted. The client (whom we had previously worked for) was advised to go to another surveyor which he did.

> The client wanted me to tell Council that works had been done when they hadn't in order that a compliance certificate be obtained. I simply refused and suggested he look elsewhere in the future.

18.3.2 Manipulation/coercion

This describes situations where people are required to participate in various unprofessional activities against their will.

While attempts at manipulation or coercion by some clients should not cause too much surprise, when the influence comes from within the profession it is rather more insidious and worrying:

> Being encouraged, when under the direction of a senior surveyor, to take a short cut which I considered to be wrong. Resolved this by going to see the Land Transfer Surveyor and asking for his advice.

And, when one surveyor had to cope with pressures from both sides in concert, the final outcome was not something he felt especially comfortable about:

> Pressure was exerted by the client and subsequently by the senior partner in the firm to lodge a Land Transfer plan before the necessary fieldwork was completed. In this case I conformed and completed the work as soon as possible. Having since considered the implications, I do not intend to repeat the practice.

A similar experience, and one equally worrying for those who are concerned about the integrity of a survey system was that:

> On one occasion I was instructed by my boss to leave an incorrectly placed boundary mark as placed instead of correcting it. My conscience was 'partially' satisfied by showing the incorrect information on the plan.

18.3.3 Technical incompetence (ineptness)

This occurs when there is a lack of skill, knowledge or ability to select or implement the appropriate courses of action.

It was noticeable in the many responses on ethical dilemmas faced, that there were very few dealing with technical issues, although one rather shame-faced respondent reported on the personal consequences of not connecting a survey properly to the survey system as follows:

> I didn't observe my origin. . . . The Department of Lands and Survey and the Survey Board made a meal out of me.

In a slightly different context, a number of surveyors related the dilemmas which they feel they face in terms of adhering to the gazetted survey regulations:

> The very frequent request to carry out redefinition surveys without lodging plans as required by survey regulations. I explain that redefinition plans are an important part of the survey system. Probably the greatest resolution of the situation is that the clients go to some other surveyor who was not concerned to lodge plans.

On occasions, they have felt sufficiently strongly on such issues not to comply with the regulations on the redefinition of boundaries. Given that the cost of doing the survey could double when a formal survey plan is required by regulation, some surveyors have jibbed:

> In redefinition surveys, the regulations would require that a plan be prepared. However, I cannot in conscience charge a client for something I believe is unnecessary.

While not technical in nature, nor part of any collusive or manipulative regime, the matter of fees and charges came into play as a dilemma for some surveyors when considering the alternative way of coping with difficult situations. This was especially true when the New Zealand surveying profession was heavily regulated and anyone cutting fees and/or charging less than the stated tariff would be dealt with severely. Some interesting tactics were employed in this regard:

> When a strict scale of charges applied, I think most surveyors experienced times when the final account, although accurate, was excessive for the work undertaken or the client's ability to pay. The accepted practice was to forward the account for the full fee but note thereon that a reduced amount would be accepted in full settlement.

And, one suspects that any surveyor who has been in a professional practice will identify with the following scenario:

> The situation arose from a small subdivision in the downtown area. A fixed price range was given and that figure was reached when the job was only half completed. The quandary was whether to submit a deficient survey plan or complete the job as it should have been. I resolved it by completing the job properly and taking a bath on profit.

18.4 PART 3: WHEN IN ROME (OR WHEREVER ELSE) . . . A CASE OF BRIBERY!

This final section sets out to examine how New Zealand professional surveyors consider an issue involving unprofessional behaviour, in this case the matter of bribery.

Interestingly, while the question of bribery is not referred to directly in the NZIS Members' Handbook referred to earlier, mention is made of the acceptance of favours by requiring that:

> All members shall avoid placing themselves under any improper obligation to any person or persons in their services on behalf of their client or employer.

For most observers, a key word in this rule would probably be the word 'improper'. It would seem reasonable to assume that most professional surveyors in New Zealand would place the acceptance or payment of a bribe, in whatever form and to whoever, in this particular category of unprofessional and/or unethical behaviour. However, later in the same listing, there is another guideline entitled *Overseas Work* which proceeds to proffer the advice that:

> In connection with work outside of New Zealand where accepted professional practices may differ from these Guidelines members may adopt local practices but shall keep in mind the principles contained in these Guidelines. Rule 20 (Code of Ethics) shall be observed by members at all times.

Unfortunately no further details are provided as to what particular form these 'local' practices which **may** be adopted might take but it is not unreasonable to assume that the possibility of having to deal with bribery was included.

Finally, when setting out guidelines on fees, the point is made that:

> No member shall offer inducements to or accept inducements from any person interested in the work undertaken by the member.

Given all this, what should a surveyor actually do when overseas and faced by a situation involving bribery. A random sample (98 surveyors) of the respondents were provided with a hypothetical scenario describing an ethical dilemma purportedly involving a New Zealand surveyor working outside New Zealand as follows:

> A New Zealand registered professional surveyor has been sent overseas by the company to undertake the survey and construction of a sizeable irrigation scheme worth some several million dollars. However, when it comes to getting things done, there are ongoing difficulties with permits for materials and labour, including bringing in much-needed New Zealand specialists.
>
> The local 'fixer', who has been seconded by the provincial Ministry of Development office, then informs the surveyor that everything will flow much more smoothly if an under-the-counter payment of approximately NZ$ 5,000 were to be made to a senior official in the local office of the Ministry. Moreover, if the money is not forthcoming, the delays will grow increasingly longer.
>
> The surveyor is not at all happy about this and makes enquiries with the following results. First, the fixer has reported the situation correctly and the official, who does effectively control such permits, does want the payment. In addition, the official is related through marriage to the Minister for Development and also the Minister of Immigration. Offering such inducements is also viewed quaintly as a local custom. However, whilst it is illegal to offer and accept such monies, the 'fixer' assures the surveyor that no-one has been prosecuted in recent memory. However, there is another complication in that recent changes in New Zealand law now make it an offence in New Zealand for any New Zealand organisation to make any such payments to overseas interests'

> Being the first overseas thrust by the New Zealand company, they had not fully appreciated the possibility of delays through such causes and there is no company policy on matters such as ex-gratia payments of this type. The surveyor is in charge of the entire operation including the payment of all normal outgoing monies although ultimately responsible to the two principals of the firm back in New Zealand.
>
> Finally, the telephone facilities are atrocious, the return airmail takes six weeks and the official now wants the payment within a week or he will make things very much worse. Oh yes, the irrigation scheme is in a very poor part of the country and is necessary to obviate a shortage of water to peasants who died in large numbers from starvation last time there were droughts. An El Nino cycle which will inevitably bring shortage of rain is forecast in the medium-term of three-to-five years by which time, barring delays, the scheme will be operational. After careful thought and some considerable anguish the surveyor decides to pay the bribe.

As can be seen, the vignette closes with the surveyor deciding to pay a bribe in order to ensure that a construction project can proceed. The surveyors were then asked to indicate whether they felt this was a correct or inappropriate action across a number of Likert-type seven-point normative scales essentially constructed on earlier work by Reidenbach and Robin with the following instructions:

> With respect to this scenario, please could you indicate your personal beliefs as to the correctness or inappropriateness of the final action which was actually taken. There are no 'right' or 'wrong' answers. (Please be frank, your confidentiality is totally assured) (Circle a suitable number for *each* of the following descriptions.)

Table 18.1 A semantic differential for ethical standards

In my view the final action taken was:								
Unfair	1	2	3	4 undecided	5	6	7	Fair
Unjust	1	2	3	4 undecided	5	6	7	Just
Culturally unacceptable	1	2	3	4 undecided	5	6	7	Culturally acceptable
Selfish	1	2	3	4 undecided	5	6	7	Unselfish

Table 18.1 Continued

In my view the final action taken was:								
Not in firm's best interests	1	2	3	4 undecided	5	6	7	In firm's best interests
Morally wrong	1	2	3	4 undecided	5	6	7	Morally right
Produced most benefit	1	2	3	4 undecided	5	6	7	Produces least benefit
Inefficient	1	2	3	4 undecided	5	6	7	Efficient

Responses on the **justice** scales (Justice, Fairness) show an interesting response to this decision in that 65% of the sample regarded the decision as unjust (responses 1 to 3) with just 20% (responses 5 to 7) seeing the decision as just. (As with other sets of responses, the middle category of 4 is assumed to be neutral here.) The responses in the fairness category are more evenly divided with 35% adjudging the surveyor's action as unfair and with 48% taking the contrary view.

The **relativist** scales (Cultural and Personal acceptability) of the decision again provide a sharp contrast, with a very strong sense being expressed that the decision, as made, is personally unacceptable to a large proportion (75%) of the group. Over half (53%) of the respondents effectively registering strong disapproval (response of 7). (Unfortunately, the 'Culturally acceptable' category is ambiguous and could have been misinterpreted by some.)

Moving to the **egoism** scales (Selfishness, In the firm's best interests), we find that overall the majority of the sample do seem to regard the decision as being more in the firm's interests (66%) than against such interests (23%). Equally the action itself is seen by more respondents as being unselfish (54%) than selfish (20%) with a large group (26%) not prepared to commit themselves.

By contrast when it comes to the **deontological** scale (Morality), a significant group (86%) consider the decision to be morally wrong and, of these, 64% take the high ground of treating it pretty much as a case of moral turpitude (response of 1).

Finally there are the two **utilitarian** scales where the responses are reasonably divided as to whether the surveyor has made a good or bad decision in this case. In a more cold-blooded approach, perhaps, his efficiency is well supported with 75% regarding the action as efficient (responses 5 to 7) and only 13% taking an opposing view. In a classic utilitarian approach of weighing up the overall balance of benefits, there is a strong consensus (79%) that there are sound benefits to be gained and a significant proportion of 41% (response of 1) essentially maintains that such benefits materially outweigh the costs.

18.5 CONCLUSIONS

The purpose of this exploratory study was, quite simply, one of exposing some of the real-life experiences and perceptions of a group of surveying practitioners. Clearly, the New Zealand Institute of Surveyors takes the matter of professional ethics and codes of conduct seriously and seeks to set out guidelines for appropriate behaviour. However, some of the behaviour required by practitioners, as detailed in the rules, could be contradictory and confusing in some circumstances.

Moreover, some of the situations with which we have been regaled, together with the diversity of views expressed in relation to a hypothetical case of bribery, suggests that these matters can be rather more complicated to unravel as and when they come to the surface. It could also be argued that the significant differences of opinion across the assessment scales, each of which is essentially built on a valid ethical approach, might actually undermine the formal rules of the NZIS anyway.

18.6 REFERENCES AND FURTHER READING

Hoogsteden, C.C. (1994a) *Ethics Education for Tomorrow's Professional Surveyors*, Commission 2 of the XXth International Congress of Surveyors (FIG), Melbourne, Australia, March, pp. 96–105.

Hoogsteden, C.C. (1994b) *The Ethical Attitudes of Professional Surveyors in New Zealand*, Commission 1 of the XXth International Congress of Surveyors (FIG), Melbourne, Australia, March, pp. 62–73.

New Zealand Institute of Surveyors (NZIS) (1994) *Members' Handbook*, Wellington, New Zealand.

Reidenbach, E.R. and Robin, D.P. (1988) Some initial steps towards improving the measurement of ethical evaluations of marketing activities. *Journal of Business Ethics*, 7, 871–9.

Sayers, D. (1931) *Have his Carcase*, Macmillan, London.

Simpson, K.W. and Sweeney, G.M.J. (1973) *The Land Surveyor and the Law*, University of Natal Press, Pietermaritzburg, Republic of South Africa.

White, L.P. and Rhodeback, M.J. (1992) Ethical dilemmas in organisational development: a cross-cultural analysis. *Journal of Business Ethics*, **11**, 663–70.

APPENDIX 18A

(Taken from the New Zealand Institute of Surveyors Rules, 29 November, 1994)

Rule 21.2 Without in any degree limiting the term unprofessional conduct for the guidance of members professional conduct shall include:

21.2.1 *Integrity and Fidelity*:

In their professional undertakings members shall:

(a) avoid placing themselves under any improper obligation;
(b) refuse to accept any reward which cannot be publicly acknowledged;
(c) respect the confidentiality of information which may be valuable or sensitive; and
(d) recognize their own professional or technical limitations or inexperience and shall at all times act in a manner appropriate to the circumstances.

21.2.2 *Continuing Education:*

All members shall maintain and shall strive to improve their competence by attention to developments relevant to their professional, technical or management activities and shall avail themselves of opportunities to further their education in those areas relevant to their activities.

21.2.3 *Employment*:

No member shall attempt to obtain employment, professional engagements or advancement by the adverse criticism of another member or other members or by any improper activity.

21.2.4 *Cooperation*:

All members, as circumstances allow, shall promote the profession of surveying by:

(a) exchanging relevant information and experience with fellow members;
(b) contributing to the work of the Institute;
(c) contributing to technical and professional publications.

21.2.5 *Education*:

All members, where appropriate, shall make their experience and expertise available to staff under their direction particularly to facilitate staff requirements for formal and practical training.

21.2.6 *Overseas work*:

All members undertaking professional engagements overseas shall uphold the ethical standards indicated in Rule 20, and abide by the established local professional practises as appropriate.

21.2.7 *Fees*:

No member shall review the fee proposed or charged by another member without the knowledge of such member.

CHAPTER 19

Case Study 2: the perceptions of ethics in practice

19.1 INTRODUCTION

Retaining professional or personal ethical standards can become increasingly difficult where, with a view toward continuation of employment or career advancement, there is temptation on an individual to attempt to improve the firm's position or meet employer expectations. Contractors are commercial, profit-driven units and profit enhancement is a primary function of surveyors employed by contracting organizations. Although this position may not be so obviously displayed in a private practice, they too now obtain the majority of their work by competition and costs must be contained within budget.

This case study focuses on the quantity surveying profession, who are employed in a range of types of employment from the traditional private practice through local authorities, utilities and housing associations to contracting and subcontracting firms. In any examination of the functions performed, such as the 'Core Skills' report in 1993 by the Royal Institution of Chartered Surveyors, the range of skills applied tends to be largely common across types of employment (although the emphasis of application of those skills can be very different). In the context of this increasingly wide range of employment type, quantity surveyors may encounter situations where their professional or personal code of ethics is challenged. This challenge may emanate from commercial drive inherent in the nature of the employment or more directly from line management pressure.

The primary professional body to which most quantity surveyors belong is the RICS. The RICS requires that its members comply with the institution's bye-laws and regulations. These bye-laws and regulations are prescriptive about minimum ethical standards and the disciplinary powers which the institution has and can apply.

Notwithstanding the institutional position, surveyors may have self-perceptions about their role within their employment and in the context of generally acceptable behavioural and ethical standards.

This chapter therefore addresses the legal, moral and institutional position and the standards displayed by experienced undergraduate students in the context of their own self-perception and their actual experience. It develops a position statement which can be tested by further research and compared with legal and institutional standards.

19.2 METHODOLOGY

A literature review was carried out to identify the broad position of criminal law as it affects professional behaviour and civil law as it influences tortious liability.

The Handbook for Chartered Surveyors was reviewed to establish the impact of the RICS bye-laws and regulations upon its members in the context of professional behaviour.

An anonymous questionnaire survey was carried out of a significant but non-representative sample of undergraduate students at the latter end of a BSc (Hons) degree course in quantity surveying and included students studying by sandwich, full-time and part-time modes of attendance. Student opinion in the context of their experience and observation by types of employment was extracted from this survey and compared with legal and institutional positions.

The limited time available resulted in the sample being undergraduates from one University. Although relatively few of the students in their responses indicated that some of their values came from their undergraduate studies, in this context it is possible that there may be a level of bias in the values expressed.

19.3 THE INSTITUTIONAL POSITION

This section is based upon the contents of *A Handbook for Chartered Surveyors – Second Edition, Professional Conduct,* by Richard Chalkley, published by the Royal Institution of Chartered Surveyors. Chalkley quotes The Ormrod Committee Report on Legal Education of 1971:

> A profession involves a particular kind of relationship with clients, or patients, arising from the complexity of the subject matter which deprives the client of the ability to make informed judgements for himself and so renders him to a large extent dependant upon the professional man. A self imposed code of professional ethics is intended to correct the imbalance in the relationship between the professional man and his client and resolve the inevitable conflicts between the interests of the client and the professional man or of the community at large.

Chalkley indicates that it is the responsibility of each member of a profession to uphold his or her professional code and that failure to do so damages the reputation of the whole profession.

The institutional bye-laws and regulations apply to all chartered surveyors (and to student and probationary members). For our purposes of this paper the example of Bye-Law 24(1) is used to illustrate the institutional position with respect to the behaviour of chartered surveyors, and in particular Section 1.15 of that Bye-Law – 'Conduct Unbefitting'.

> 'No member shall conduct himself in a manner unbefitting a chartered surveyor'.

This bye-law governs conduct which is considered to be inconsistent with a chartered surveyor's membership of the Institution.

Chalkley has identified among others the following as recent illustrative examples of contravention of this bye-law:

1. Failure to pay moneys due under a court judgement.
2. The misappropriation of money.
3. The writing of a rude and offensive letter to a former client.
4. A breach of client confidentiality.
5. Failure or delay in replying to correspondence from a client.
6. Being convicted of a criminal offence involving dishonesty.
7. Being adjudged bankrupt.
8. Failing to discharge clients' instructions promptly, having accepted them.
9. Permitting a candidate to cheat in a Test of Professional Competence.

There are bye-laws and regulations relating to, among other issues, the accuracy of publicity and Continuing Professional Development and these should be referred to to obtain a comprehensive view. The bye-laws and regulations are less comprehensive in terms of chartered surveyors working outside private practice and relate largely to nameboards and disclosure.

19.4 THE LEGAL POSITION

19.4.1 Criminal law

Bye-law 25(2) gives the General Council of the Institution the power to expel a member summarily or to refer a matter to the Disciplinary Board. This power can be used where a member has been convicted of a criminal offence involving 'embezzlement, theft, corruption, fraud or dishonesty of any kind or any other criminal offence carrying on first conviction a maximum sentence of not less than 12 months imprisonment'. This is the Institutional position, but it is of course by the nature of things the ethical position imposed by the legal system current in the UK at the time of the relevant action.

19.4.2 Civil law

Surveyors can be held liable under the civil law relating to contract and to tort. It is not our intention to explore these issues in any great depth but practising surveyors are expected to meet the requirements of their contract and of the law of tort in the carrying out of their duty. They can become liable in this context to their client, their employer, their employees, their insurers, and the general public.

The Institution has well-monitored rules about professional insurance to protect clients and all employers must have third-party insurance.

In drawing together the legal and Institutional requirements and expectations a view can be developed which is consistent with Chalkley's remarks about the Ormrod Committee Report and each member's responsibility. In some ways the position of those surveyors employed in contracting organizations seems insufficiently addressed and of those in private practice over prescriptive.

The interpretation of the limited literature review seemed to suggest a position which can be used as a standard in terms of the expected standard of behaviour of a chartered surveyor. In itself this is considered to be incomplete because chartered surveyors are human beings with ethical standards which may be related to given values, developed values, religious beliefs, etc.

19.5 RESEARCH PROCESSES

In striving to achieve the aims of the study, the professional standard displayed and experienced by undergraduate students was identified as one way to compare and contrast the Institutional and legal position with current practise and attitudes.

The manner in which the ethical perceptions of undergraduate (and practising) quantity surveyors has and will be collected is through the medium of an anonymous questionnaire. This questionnaire was developed to facilitate an examination of the following broad areas:

- general ethical standards, attitudes and current practice
- the effects of employment type upon current practice and ethical standards
- the effects of years of experience upon ethical standards and attitudes.

In the development of the questionnaire a number of factors were addressed. These included:

- the need to obtain statements about current practice (factual answers);
- the need to obtain responses about what practising quantity surveyors have experienced or observed; and
- the need to obtain opinions or views about issues relating to practice.

In some cases questions were intentionally linked and in some cases they have been carefully positioned in the questionnaire to avoid leading questions. In most cases the questions require a closed response to facilitate analysis.

The questionnaire was piloted with a representative sample of the target undergraduate responses and minor adjustments were made.

19.6 SELECTION OF RESPONDENTS

In this study, undergraduates in the final stage of the BSc (Hons) Degree in Quantity Surveying were used. The students belong to three separate cohorts namely;

- **SW4 (Sandwich 4)**: year 4 of a 4-year sandwich degree with 2 prior years of full-time study (year 1 and 2) and 1 year of workplace experience (year 3).
- **PT4 (Part-time 4)**: year 4 of a 5-year part-time degree. The part-time study mode requires approximately 35 days of University attendance per annum, the students are required to be in full-time quantity surveying employment.
- **PT5 (Part-time 5)**: year 5 of a 5-year part-time degree.

Table 19.1 provides an analysis of these respondents by age, employment type, gender and experience.

Notes

1. The average years of experience as a quantity surveyor shown in Table 19.1 reflect experience gained by some students before commencement of the course.
2. To facilitate analysis, those students who fall outside the main employment types (i.e. private practice, contractors, sub-contractors and local authority) have been excluded. The types of employment have been consolidated to two main categories: (i) private practice and local authority; (ii) contractor and sub-contractor.

Table 19.1 Analysis of respondents by age, employment type, gender and experience

	SW4	*PT4*	*PT5*	*Total*
Age				
18–21	6	3		9 (7%)
22–25	26	40	5	107 (80%)
26–28	3	5	5	13 (10%)
29–35		1	2	3 (2%)
36–40			1	1 (1%)
Over 40				
Total	35 (26%)	49 (37%)	49 (37%)	133
Employment by type				
Private practice	14	11	14	39 (29%)
Contractor	12	28	29	69 (52%)
Sub-contractor	1	4	1	6 (5%)
LA	1	3	3	7 (5%)
Others	7	3	2	12 (9%)
Total	35	49	49	133
Average years of employment as QS				
Contractor/sub-contractor	1.38	5.25	6.03	4.89
PQS/LA	1.73	4.64	5.65	4.07
Gender				
Male	33	47	46	126 (95%)
Female	2	2	3	7 (5%)
Total	35	49	49	133

19.7 ANALYSIS OF RESPONSES TO THE QUESTIONNAIRE SURVEY

With reference to the criteria outlined in the methodology, the questionnaires have been analysed in terms of the stated objectives.

19.71 Consideration of general ethical standards, attitudes and practice and the influences of employment background.

The indicators used for such assessment are:

Loyalty	(All)
Acceptance of gifts	(PQS)
Contractor perceptions	(Contractors)
Ethical values and sphere of influence	(All)
Membership of professional bodies	(All)
Levels of malpractice	(All)

Table 19.2 Attitudes to loyalty

Who do you owe loyalty to?	*Response from PQS/LA (max. 46)*		*Response from contractor/sub-contractor (max. 75)*		*Total (max. 121)*	
Employer	38	(83%)	64	(85%)	102	(84%)
Client	28	(61%)	9	(12%)	37	(31%)
Self	26	(57%)	45	(60%)	71	(59%)
Who do you owe your first loyalty to?						
Employer	25	(54%)	39	(52%)	64	(53%)
Client	4	(9%)	2	(3%)	6	(5%)
Self	17	(37%)	34	(45%)	51	(42%)
Total	46		75		121	

Where appropriate, statistics have been prepared indicating the relative influence of type of employment.

With regard to attitudes to loyalty, the statistical analysis suggests the following (Table 19.2):

1. Irrespective of employment type, most undergraduates believe they owe loyalty to their employer (84%).
2. The majority (61%) of undergraduates employed in private practice/local authority believe that they owe loyalty to the client as opposed to a minority (12%) of undergraduates employed by contractors/sub-contractors.
3. Irrespective of employment type, the majority of undergraduates believe that they owe their first priority to their employer (53%).
4. There is a belief in a majority of graduates that they owe a significant duty of loyalty to themselves.

With regard to attitudes of quantity surveyors in private practice/local authority to the acceptance of gifts from contractors, the analysis (Table 19.3) suggests the following:

1. That 91% would accept a gift from contractors.
2. A corresponding 89% believe that acceptance of a gift would not affect impartiality when dealing with a contractor.
3. Attitude to gift acceptance is greatly influenced by nature and value of the gift.
4. A substantial number of undergraduates believe that the acceptance of high value gifts would not affect their impartiality when dealing with contractors.

Table 19.3 Acceptance of gifts

Would you accept a gift from a contractor?		
Yes	42	(91%)
No	3	(7%)
Nil return	1	(2%)
Total	46	
What level of gift would you accept? (max. 46)		
Pen	39	(85%)
Bottle of whisky	40	(87%)
Case of champagne	23	(50%)
Weekend break	16	(35%)
Week's holiday	14	(30%)
Cheque	7	(15%)
Cash	7	(15%)
Would acceptance of a gift affect impartiality?		
Yes	2	(4%)
No	41	(89%)
Nil return	3	(7%)
Total	46	

With regard to the attitudes and perceptions of quantity surveyors employed by contractors, the statistical analysis (Table 19.4) suggests the following:

1. That 91% of respondents stated that the primary function of their employment was to make profit (some respondents stated both profit and cash flow) and 75% would not be concerned if their objectives disadvantaged the client.
2. A near total of respondents (96%) would enhance remeasured quantities or dayworks if asked to do so by their superior 'if they could get away with it'.
3. A large majority of respondents (85%) would attempt to unfairly beat down the price of (say) a sub-contractor by utilising technical or intellectual advantage.
4. The responses are consistent with those summarized in Table 19.2 (Attitudes to loyalty).

With regard to ethical standards, values and influences, this statistical analysis (Table 19.5) suggests the following:

1. Undergraduates employed by contractors are more likely to believe that commercialism must come before an ethical stance in 1995.
2. The large majority of undergraduates (81%) would question with a higher authority any action for which they had doubts.
3. The main influences regarding ethical values of right and wrong

Table 19.4 Contractor's attitudes and perceptions

What do you consider your primary functions? (max. 75)		
Profit	68	(91%)
Cash flow	12	(16%)
Other	2	(3%)
Would you be concerned if your objectives disadvantaged the client?		
Yes	19	(25%)
No	56	(75%)
If asked by your superior to enhance measured work or daywork would you?		
Do it if you could get away with it	72	(96%)
Never do it	1	(1%)
Other	2	(3%)
Would you put your employment at risk if asked by your superior to enhance measurement or daywork?		
Yes	6	(8%)
No	69	(92%)
Total	75	
Would you beat down the price of a sub-contractor if you had a technological/intellectual advantage?		
Yes	64	(85%)
No	11	(15%)
Total	75	

appear to be similar in both employment sectors. Undergraduates believe that the main provider of values of right and wrong are parents (84% of respondents), friends (37% of respondents), school (38% of respondents) and university (24% of respondents).

4. A large minority of undergraduates employed in the contractor sector (16%) expressed concern at the level of ethical standards they put into practice in their job.

With regard to membership of a professional body, the statistical analysis (Table 19.6) suggests the following:

1. The majority of undergraduates (72%) believe that membership of a professional body will not affect their ethical standards applied at work.
2. Approximately half of the undergraduates expressed awareness of rules of professional behaviour. The detail of such rules provided by

the respondents was generally vague. The following items were stated:

- the requirement to maintain Professional Indemnity Insurance
- restrictions on advertising

Table 19.5 Ethical standards, values and influences

Do you believe that in 1995 commercialism must come before an ethical stance?	*Response from PQS/LA (max. 46)*		*Response from contractor/sub-contractor (max. 75)*		*Total (max. 21)*	
Yes	24	(52%)	52	(69%)	76	(76%)
No	20	(43%)	21	(28%)	41	(34%)
Nil return	2	(5%)	2	(3%)	4	(3%)
Total	46		75		121	
If colleagues of equal standing to yourself confirmed that an action was acceptable although you had doubts, would you?						
Do it	1	(2%)	8	(11%)	9	(7%)
Refuse and stand alone	4	(9%)	10	(13%)	14	(12%)
Question with higher authority	41	(89%)	57	(76%)	98	(81%)
Total	46		75		121	
Where did you obtain your values of right and wrong?						
Parents	35	(76%)	67	(89%)	102	(84%)
Friends	17	(37%)	28	(37%)	45	(37%)
School	20	(43%)	26	(35%)	10	(8%)
Bible	–	–	1	(1%)	1	(1%)
Media	7	(15%)	7	(9%)	14	(12%)
University	13	(28%)	16	(21%)	29	(24%)
Other	8	(17%)	8	(11%)	16	(13%)
Total	N/A		N/A		N/A	
Do you consider your ethical standards as applied to your job are at an acceptable level?						
Yes	44	(96%)	60	(80%)	104	(86%)
No	–	–	12	(16%)	12	(10%)
Nil return	2	(4%)	3	(4%)	5	(4%)
Total	46		75		121	

N/A, not applicable

Table 19.6 Membership of professional bodies

Does or will membership of a professional body (such as RICS) affect your ethical standards at work?	*Response from PQS/LA (max. 46)*		*Response from contractor/sub-contractor (max. 75)*		*Total (max. 121)*	
Yes	13	(28%)	18	(24%)	31	(26%)
No	32	(70%)	55	(73%)	87	(72%)
Nil return	1	(2%)	2	(3%)	3	(2%)
Total	46		75		121	
Are you aware of any rules of professional behaviour?						
Yes	24	(37%)	37	(49%)	61	(50%)
No	21	(46%)	36	(48%)	57	(47%)
Nil return	1	(2%)	2	(3%)	3	(3%)
Total	46		75		121	

- prohibition of bribes and fraud
- the need for fidelity to clients

While no statistic is available, it is likely that most of the undergraduate respondents are student members of the RICS; this presumption is based upon a perceived general desire among most undergraduates to gain Chartered Surveyor status at the earliest opportunity, thus requiring to register and record their industrial training period (applicable to sandwich students) or working experience (applicable to part-time students).

With regard to malpractice, the statistical analysis (Table 19.7) suggests the following:

1. Knowledge of malpractice is more prevalent among undergraduates employed by contractors than those employed in private practice.
2. Certain malpractice appears to be endemic (e.g. measurement enhancement, fabrication/enhancement of dayworks).
3. Fraudulent activity is at a relatively high level.
4. Bribery payments and threats to persons in their employment appear to be more than exceptional and rare occurrences.

19.7.2 Consideration of the effects of years of experience upon ethical standards and attitudes

A comparison of the responses from undergraduates in the Sandwich 4 cohort (average years of employment 1.38/1.73) with those from the

Table 19.7 Levels of malpractice

Have you been aware of the existence of the following in your working experience?	*Response from PQS/LA (max. 46)*		*Response from contractor/sub-contractor (max. 75)*		*Total (max. 121)*	
Measurement enhancement	34	(74%)	68	(91%)	102	(84%)
Concealment of faulty work	14	(30%)	41	(55%)	55	(45%)
Substitutions of below specification work/ material	23	(28%)	34	(45%)	47	(39%)
Fabrication/enhancement of daywork	23	(50%)	65	(87%)	88	(73%)
Photocopy letter heads/ invoices to fabricate records	10	(22%)	39	(52%)	49	(40%)
Action to unfairly bankrupt for gain	5	(11%)	12	(16%)	17	(14%)
Excessive hospitality	18	(39%)	29	(39%)	47	(39%)
Favours promised for favours given	10	(22%)	29	(39%)	39	(32%)
Bribery payment	1	(2%)	17	(23%)	81	(15%)
Threats to persons or their employment	5	(11%)	18	(24%)	23	(19%)
Total	133		352		485	
Overall 'response rate'	29%		47%		40%	
Is it ever ethically justifiable to offer some sort of bribe or excessive hospitality to obtain a contract?						
Yes	10	(22%)	31	(41%)	41	(34%)
No	32	(70%)	40	(53%)	72	(72%)
Nil return	4	(8%)	4	(6%)	8	(6%)
Total	46		75		121	

PT4/PT5 cohorts (average years of employment 5.25/4.64 and 6.03/5.65 respectively) has been made. This comparison revealed little variation between the groups (distinct in terms of experience of Quantity Surveyor Profession/Construction Industry) and meaningful comment as to the effect of years of experience within the profession has not been possible.

19.8 CONCLUSION

In testing undergraduate opinion certain expectations were developed before any results were recovered. These expectations included; a difference between Institutional (RICS) requirements and current practice and opinion; a lowering of standards as perceived by practitioners falling into the 'older than 35' category, or 'older than 40' category.

The latter expectation remains inadequately tested at this stage and will be considered further as the research develops. The former expectation has been discovered to be correct, in terms of this sample, to quite an extreme extent. This is particularly highlighted in the responses which related to the acceptance of gifts by private practice and local authority quantity surveyors and to the general area of 'malpractice'. In some cases the gifts which would be accepted would by any standard be considered to be generous, although in the majority of cases, the respondents believed that their impartiality would not be threatened.

In terms of 'malpractice', certain types seem to be endemic, with types of fraud at an alarming level and bribery payments not uncommon.

Sadly, the majority of respondents believed that membership of a professional body would not affect their ethical standards as they are applied at work, although a significant minority of those employed by contractors are themselves dissatisfied with the ethical standards applied to their job.

Significant responses were obtained relating to opinion about loyalty and commercialism, with most feeling that first loyalty was due to the employer and that commercialism must come before an ethical stance. This is in stark contrast to the quotation from the Ormrod Committee offered by Chalkley (1994) and stated above whereby the client is perceived as being deprived of the ability to make informed judgements for himself and dependent upon the professional man or woman.

It is somewhat heartening to see that most private practice and local authority respondents felt some loyalty to the client, although little evidence of such loyalty is apparent from the responses provided by respondents from the contracting sector.

19.8.1 Suggested position statement

At this point in the research it is possible to develop a position statement while acknowledging that this will change over time. A suggested position statement at this time is as follows:

> There is a significant variance between perceived and/or expected standards and real standards in the practice of quantity surveying.

Notwithstanding the understood ethical position vis-à-vis the chartered quantity surveyor and his/her client, first loyalties are generally seen to be due to employers, and commercialism is seen to be of primary importance. There is evidence of differing attitudes to ethical issues between private practice/local authority respondents and contractor/sub-contractor respondents. There is evidence that malpractice, fraud and bribery exist to some extent. Membership of a professional body is not seen to be significant in terms of applied ethical standards.

CHAPTER 20

Defining service quality and the merits of introducing a formal system

20.1 INTRODUCTION

Little evidence exists to support the claimed benefits relating to certified quality management systems in professional firms. This chapter details one aspect of a research study undertaken by the Construction Industry Research and Information Association (CIRIA) to survey the experiences of firms who have implemented systems to comply with BS 5750. The findings reported here are concerned with the measurement of benefits. While the research has yielded few clear-cut conclusions, it has provided insights into particular areas of management systems which firms believe have provided benefits.

We will focus mainly on the UK's approach to quality management epitomized by British Standard 5750. The material is relevant, coming at a time when both proponents and customers are unable to agree as to whether or not its implementation has yielded the tangible benefits that its champions promised.

In particular, we will concentrate on a broad cross-section of firms operating within, and serving the needs of the construction industry, and is taken from a research project undertaken for the CIRIA by Bucknall Austin plc and Liverpool John Moores University. It is emphasized that the report is principally a **survey** – to enable the reader to form a general idea of the arrangement and chief features of quality management systems (QMS), and it is based upon experiences – i.e. actual observations or practical acquaintances with the facts and the events.

The survey was designed in three parts. The first (Part A) was undertaken in 35 organizations representing a broad cross-section of the construction industry. It is the results of this part of the exercise which are reported below. The second part involved 30 client organizations, and the final part sought to obtain information from firms who had considered and consciously rejected the implementation of certified management systems.

We will discuss the methodology used in the research, which obtained both quantitative and qualitative data, based around a framework of eight principal criteria used to measure the impact of BS 5750 on the management of organizations which have implemented a quality management system based upon its requirements.

The analysis of the results provides indications of the extent to which benefits have been realized.

The later sections focus on firms looking towards quality accreditation, and the implementation of procedures compatible with the requirements of BS 5750/ISO 9000. Drawing on recent research including CIRIA and the RICS, we will attempt to identify some of the more fundamental issues affecting a firm's capacity to make an effective transition to quality management. As a conclusion, we question whether certification should necessarily be the first step into quality management.

20.2 THE MERITS OF FORMALIZED QUALITY MANAGEMENT SYSTEMS

20.2.1 Methodology

The research method was designed in consultation with a steering group established by CIRIA. It was a requirement of this group that firms taking part in the survey would have been certified for a minimum period of 2 years. One of the reasons for this requirement was the belief that the survey should record experiences of firms operating with systems which had been well 'bedded in'. Thus, it was important that readers of the final report would be able to assess the likely effect of a system in full operation, rather than one which was in the process of implementation, or experiencing early teething problems.

Part A was undertaken in 35 organizations representing a broad cross-section of the construction industry. Wherever possible, interviews were conducted with three people selected from senior and middle management, and a more junior member of staff or operative. The intention was to obtain opinions from both implementers and users of the QMS. People were questioned in private, using a detailed questionnaire which had been sent to the organization in advance of the meeting. A total of 87 people were interviewed and approximately 30 000 items of data collected. The survey was undertaken by three researchers over a period of 6 months commencing in October 1992.

The nature of BS 5750 is significant in the flexibility it affords the user in designing a QMS which is most suited to their particular organization. However, this presents a serious difficulty in terms of how the effects of changes are identified and measured. A simple approach could monitor profitability both pre- and post-implementation, although this would present difficulties in terms of distinguishing between factors which were attributable to a QMS and those which were not. Therefore, a more discriminating approach was required.

The research took the form of interviews with managers in firms who had implemented systems to comply with BS 5750. Eight principal criteria were used to measure the effect of a QMS upon an organization, which in turn were developed into a series of 24 statements which could be used in face to face interviews (Table 20.1). The firms included professional services organizations (i.e. architects, consulting engineers, quantity surveyors), contractors (main and sub-contract) as well as manufacturers and suppliers.

To measure the benefits that the QMS had brought to the organization, interviewees were asked to indicate the strength of their agreement with each of the statements in Table 20.1. This information was recorded on a numeric scale as depicted in Table 20.2. Each piece of paired data provides three insights as follows:

1. Expectations (E) – how high were they?
2. Perceptions (P) – how high were they; in addition, whether the respondent felt that a beneficial change had occurred (a score of 5 or greater; 4 being the neutral mid-point).
3. Level of satisfaction – by calculating $P_n - E_n$ the following conclusions can be drawn:

   - a positive result indicated that the QMS had satisfied the individual's requirements, that is, what they have received as a benefit is greater than they had expected, the magnitude of the result indicates the strength of their satisfaction.
   - a negative result indicated that expectations had not been met; this should not be taken to indicate an adverse effect however, merely one that has not been as advantageous as had been expected.

20.2.2 Findings

The findings reported in this paper are limited to Part A of the survey, as previously indicated, and the results of the first interviews conducted with people responsible for implementing the QMS.

The findings are reported at two main levels of analysis. Firstly, an

Table 20.1 Measurement criteria and evaluative statements

Measurement criteria	*Evaluative statement*
Company image The system has:	1. Enhance the firm's image
Industry image BS 5750 has:	2. Enhance the image of the construction industry
Productivity The system has:	3. Improved productivity throughout the firm 4. Improved information flow 5. Has reduced the time it takes to deal with queries 6. Clarified responsibilities 7. Reduced the amount of paperwork
Cost savings The system has:	8. Resulted in a reduction in administration costs 9. Achieved savings through a reduction in errors/failures 10. Achieved savings through a reduction in remedial work
Certainty The system has:	11. Improved accuracy in predicting future activities 12. Improved control of resources 13. Increased certainty of achieving deadlines 14. Given greater confidence that targets will be achieved
Morale The system has:	15. Improved my job satisfaction 16. Improved satisfaction within the firm
General management The system has:	17. Improved the identification of management issues 18. Improved operational matters – your projects/products 19. Resulted in less management attention required for routine matters 20. Resulted in a reduction in the amount of crisis management
Committed customers The system has:	21. Resulted in greater client/customer satisfaction 22. Resulted in a greater number of opportunities 23. Increased sales 24. Resulted in an increase in repeat business

overall impression in terms of whether changes arising out of implementing a QMS have been beneficial, and to the extent expected; secondly a more detailed examination of the eight criteria adopted.

An analysis of the overall results, taking 4 as neutral, indicates that

Table 20.2 Mechanism to record the benefits of implementing a QMS

Expectations at the outset							*Perceptions of the benefits which have been achieved*						
Strongly agree						Strongly disagree	Strongly agree						Strongly disagree
7	6	5	4	3	2	1	7	6	5	4	3	2	1

the mean level of expected benefit from all respondents was 4.6, whereas the perceived benefit amounted to 4.3. Thus, while the general effect of implementation had shown a slight benefit it can be stated that managers were a little disappointed with the overall effect. This is indicated by the mean expected effect pre-implementation of 4.6, marginally greater than that achieved post-implementation. In other words, their systems were not living up to expectations.

Whereas in overall terms the sample has indicated at best an indifferent response to BS 5750, this disguises a more emphatic picture for individual firms and particular criteria. The statistical analysis has been embellished with anecdotal evidence in the form of comments made by interviewees during the assessment exercise.

Company image

Some 73% of managers indicated that the image of their firm had been enhanced as a result of implementing BS 5750:

- 'creates the impression of a club'
- 'name more widely known, confirmed our good reputation'
- 'work gained from high-profile clients'

Industry image

Only 26% felt that this had been enhanced, therefore a majority of people believed that the standard has had no effect:

- (BS 5750 is) 'expected to clear out the scruffy end of the market'
- (we have) 'increased confidence when dealing with registered firms'
- (there is the) 'traditional conservatism of the industry, and a lack of client demand'

Productivity

A majority indicated that productivity had improved. This was mainly attributed to improved information flow (74% of managers) and clarified responsibilities (74%):

- 'increased the amount of paperwork, for the better!'
- 'has focused more attention on the sites, away from the head-office'
- (has resulted in) 'clearer operating procedures

Cost savings

The examination of any savings in costs revealed that half of the managers felt that this had been achieved, although there had been an increase in the cost of administration which was offsetting some of the savings. Some 59% indicated savings through a reduction in errors/failures, and 41% felt that they had been successful through a reduction in remedial work.

Certainty

The fifth criterion, certainty, was intended to test the extent to which the range of operations in each organization had been brought under control as a result of implementing BS 5750; 53% of respondents were generally of the opinion that this had been achieved:

- (the system) 'instils confidence – often misplaced!'

Morale

An attempt was made to record changes in morale, since it has been suggested that people prefer to work for companies which have a reputation for quality. The introduction of a QMS was felt to have benefited organizations in terms of both the morale of management and generally within the firm, although an almost equal number of respondents indicated that there had been no change:

- 'we can now work more as a team, training men to do other jobs to help us achieve our deadlines'
- 'morale has decreased because of all the checking'
- (there is) 'a lack of commitment from upper management

General management

Benefits in terms of improved identification of management issues and operational matters was successfully recorded by 70% of managers. The management of routine matters, and crisis management was felt to have been influenced by half of those surveyed:

- (BS 5750) 'has prompted and promoted discussion of issues which are not necessarily quality related'

- 'it encourages people to work with the company, the system is not there to catch you out'
- 'generally improved management's awareness of problems'

Committed customers

The research for this criteria looked at four issues: client/customer satisfaction, number of opportunities, increased sales and repeat business. Some 70% of managers reported an increase in opportunities following certification, whereas the other three statements recorded a less marked level of agreement, with between 38% and 56% of respondents stating that their QMS had provided the benefits listed in the questionnaire:

- 'external factors (to the QMS) are prevalent, price dictates success'
- 'a new section has been added to the quality manual dealing with customer complaints'
- 'as a result of the procedures, a feeling that clients are more appreciative of the way things are handled'.

The detailed findings can be summarized at the level of the measurement criteria in terms of whether or not the introduction of a QMS has brought benefits to the firms, and if firms are satisfied with the result. The findings are summarized in Table 20.3.

Table 20.3 Summary of benefits and satisfaction of a QMS

Evaluative criteria	*Overall benefit achieved*	*Satisfaction with result*
Company image	Yes	No
Industry image	No	No
Productivity	Yes	No
Cost savings	No	Yes/No
Certainty	Yes	No
Morale	Yes	No
General management	Yes	No
Clients/customers	Yes	No

20.2.3 Conclusions

Before commenting upon the detailed findings it is necessary to make a number of general points about the research methodology and the general business environment at the time the data were obtained. The rationale behind the requirement that a firm should have been certified for a minimum period of 2 years has already been outlined. While this was thought to provide a more relevant source of experiential insight for firms considering BS 5750, it must also be considered that another effect would be to limit the choice to 'pioneers' in the field – in other words, firms which had embarked on the introduction of a management system at a time when there were very few in the construction industry who had experience of the standard. This was also reflected in the shortage of experience of the certification bodies who were attempting to advise these firms. One of the likely effects is that an organization looking to implement a formal QMS more recently is likely to find more advice available. Thus, the resources required to introduce and maintain the system would be less.

Secondly, the survey was conducted at a time of recession in the industry, and there was often a clear sense of frustration among individuals that their efforts were often thwarted by purchasers whose sole criterion appeared to be price.

On the whole the research has provided few clear-cut conclusions. There are perhaps a number of reasons for this, not least of which is the diverse nature of the organizations who participated in the survey. In addition, it would seem that one of the major factors influencing the degree to which a QMS will benefit a firm, is the congruence and salience of pre-existent managerial mechanisms, before the introduction of the certification process to BS 5750. Furthermore, the research has illustrated the importance of defining clear objectives before embarking on the process of implementing a QMS, and being able to measure its cost and effectiveness.

20.3 FACTORS AFFECTING A FIRM'S ABILITY TO MAKE A SUCCESSFUL TRANSITION TO QUALITY MANAGEMENT

20.3.1 Introduction

This section focuses on the approach to QM expressed in compliance with BS 5750/ISO 9000, a 'horizontal' standard, which provides a framework for the introduction of a quality policy. This particular quality initiative is claimed to be the most extensive initiative directed to quality in the whole of Europe, but which has experienced much lower adoption rates in service-based organizations than had been anticipated.

By examining the research work carried out to date on this subject, doubts have been revealed about the effectiveness of the change to the standard in producing results which are congruent with original aspirations. Some employees have not been fully convinced of the success of making the transition in their firms from the position where they were, to one governed by management and administrative procedures appropriate to meet the requirements of BS 5750. However, there are at least some firms who appear to be more satisfied than others by the results of making this change. We will try to look at some of the key issues which firms should address if they are to make a successful transition to quality management.

How can we recognize which firms may be better at effecting the change to quality management? The 1995 CIRIA survey helps us to identify firms in different sectors of the construction industry which gave consistently better results in matching employees' expectations of the benefits accruing, with the perceived reality after the event, when measured against a number of principal criteria. Analysis of the CIRIA report reveals that results varied considerably both between firms and sectors. The less-than-positive outcomes of the change to quality management systems, may have been attributed to insufficient focus on those factors impacting upon the management of the processes and procedures effected during the transition phase.

Significantly, given the impact of the changes on the organizations, initial steps in the process were characterized by a lack of preparation, with the report indicating that one quarter of the firms surveyed had no clear methodology for the implementation of their QMS. Their state of preparedness before commencement was limited and in some cases insufficient thought had been devoted to the development of a clear and coherent strategy for effecting the transition from one state to another. Drawing from other reports by the RICS, Binney and CIRIA, a wide range of factors can be seen to impact on the process of transition. It was clear that many firms failed to understand the relative complexity of the process they were undertaking, and although the initial outcome of certification to the standard had been superficially successful, the consequent price paid in terms of its impact on employees, was heavy.

20.3.2 Self-assessment

The RIBA report of 1990 highlighted the importance of self-assessment before embarking on the route to quality management, while the RICS report of 1992 drew attention to the fact that many firms failed to take this vital step. Self-assessment requires a detailed examination of those factors, both internal and external, which define the fabric of the firm, e.g. the client base, the efficacy of the organization structure to support the changes, the scope of activities to be included in certification, etc. When undertaking this process, there may be a temptation to perpetuate existing traditional approaches, at the expense of more efficient and appropriate methods. CIRIA addressed this issue, focusing on the extent to which amendments and changes in documentation and procedures had been a necessary precursor to transition to quality management. It showed that for construction-related firms, an average of 70% of the procedures contained within the quality framework were likely to be amendments of existing or wholly new procedures. Some quality-related procedures, for example, internal quality auditing, invariably were non-existent before implementation of the scheme.

20.3.3 Information and organization structure

A central feature of any scheme of quality management is the way in which the systems and procedures prescribed, facilitate information transfer. Lansley provides us with an insight into the links existing between organization structure/types and their capacity to facilitate information diffusion. Boisot's codification–diffusion theory is cited as a means of describing '. . . the information-processing differences inherent in the structures of UK contractors . . .' The distinction is made between types of information which are amenable/submissive to codification, i.e. readily capable of being encapsulated in a set of rules, and those less capable of expression in quantitative terms. Some information is seen as more readily capable of diffusion/dispersion, compared with other types which '. . . require a high level of expertise to both encode and decode. . . .' Consequently, different types of organisation can be defined, dependent on the degree to which they facilitate codification and foster diffusion of information. Hence, a firm which has adopted a structure which is amenable to a high degree of codification of information, should facilitate '. . . the development of clear expectations and standards of performance and this reduces conflict due to incongruent goals and systems of organisation'. Consequently there is the anticipation of higher levels of control and confidence that the system is working as designed.

Organization types which facilitate/foster the diffusion of information are recognized by the operation of a less integrated and a more fragmented form of organization structure. Consequently, at a fundamental level, the way in which the firm is structured can either facilitate or mitigate against the ease with which it makes the transition, depending on the type and complexity of its information. Furthermore, Lansley links firm types and classifications with the capacity of the design to facilitate control, foster well-defined systems of integration via groups/teams, promote effective formal channels of information and accessibility of information to do the job. From this, it could be concluded that firms who displayed these positive structural characteristics, are more likely to effect an easier transition to certification. One could conjecture that this is likely to be more important, the greater the magnitude of changes envisaged to existing documented procedures. In the CIRIA study, product manufacturers were highlighted as requiring a high proportion of new systems and procedures devised specifically to comply with the QMS. Of the sectors examined, architectural firms introduced the lowest proportion of new procedures (19%).

20.3.4 Senior management involvement

Binney's study in 1992, researching a much wider spectrum of industry and commerce, focused on both Total Quality Management (TQM) and Quality Management systems (QMS) installed in accord with BS 5750, echoed some of the above findings. In particular he identified the leadership issue as a key factor. To promote effective transition, Binney highlighted a number of key traits which should be displayed by those engaged in the process of leading and effecting change. He maintained that '. . . strong leadership has been present in the companies that have an effective process of change'. Such leadership has been characterized as being forthright, displaying a willingness for risk-taking to meet key objectives and a capacity to listen and involve employees in operationalising objectives. Complementary to this, was that the quality manager should be intimately conversant with mechanics of how the work is done. The more successful had established credibility at all levels within the firm, often built up over a period of years working their way through the ranks of the business. Furthermore, a high degree of persistence was considered desirable in pursuing objectives and a display of consistency in supporting and dealing with employees.

The RICS found that senior employees appointed to quality management roles were often appointed for the wrong reasons, many with a paucity of skills and limited experience for managing change. Instances were cited of senior staff '. . . nominated to undertake the implementation of the system on the basis that they had more time than others, or had once read an article about quality management'. A superficial approach will often lead to a superficial and less than satisfactory result.

In making the transition, senior management will often call upon independent quality consultants as a means of enhancing their skills. Over-reliance on such appointments can lead to problems. Care must be exercised in delegating responsibility to consultants, whose interpretation of the framework may not entirely accord with the way in which employees discharge their activities. Equally, care must be taken to ensure that emoployees do not receive the wrong signals about senior management commitment. Some employees could view the appointment, as a signal that senior management is abrogating its responsibility for effecting the transition, particularly if the consultant is allowed to become the driving force for change, with the former taking a back-seat. In which case senior management may be sending the wrong signals to employees, and in the process possibly being perceived as attempting to limit their involvement and consequent commitment.

A strong commitment to quality is likely to be insufficient, unless senior management can demonstrate sufficient knowledge and capacity to lead this process of change successfully. Some approaches demonstrated insensitivity, plunging employees straight into the process of change with little prior warning or explanation. Later, considerable effort may need to be expended, in mitigating the damage and placating unhappy staff. Such a 'sudden death' approach is bound to foster insecurity in employees and can make them unwilling contributors to the process of change. Bleer also points to the danger of senior managers regarding the introduction of quality management as a 'programme', a discrete activity, with set beginning and end points, once installed then forgotten, as senior managers target their next focus of attention. However this is unrealistic, as once started the process continues, even after certification, when the real business of maintaining, motivating and refining changes takes place. Concomitant with this, Binney identified some of the following as key features of successful implementation.

- Direction from the top but not imposition.
- Involvement of all employees.
- Helping employees to understand how the firm works and thus fostering learning and cooperation (the production of extensive written manuals to describe the processes and procedures was in some ways less important than the insight gained in developing them. This vital learning experience has been underplayed in terms of its capacity to encourage both the employees and the company to grow).
- Encouraging employees to challenge existing procedures and systems.

20.3.5 Assessing the climate for change

As part of a firm's self-assessment, it is imperative for those charged with the task of effecting the transition to certification, to gain a clear perspective of the readiness of the employees to engage in the changes likely to ensue by implementing BS 5750. Even if they do not like what they see, they may at least gain some awareness of both the magnitude and consequences of undertaking this transition and may be able to impart to employees what level of transformation they are seeking to achieve. Aspirations for the change, when conveyed to employees, need to be realistic. No attempt should be made to sell the transition to quality management as some form of religious transfiguration.

Other writers in the field seek to differentiate a number of different dimensions of change, including imperatives for change, contributory reasons for resistance, strategies and structures for implementation, the implications for individual, group and organization as well as the political dimensional. An RICS research report in 1993, which focused on quality management in the surveying profession, concluded that many of those driving the process were insufficiently imbued with the requisite level of management training and skills necessary to effectively plan and manage the transition.

The capacity of senior management to manage change and to see it through are essential elements. However, the purist view that this process should be controlled and orderly often contrasts with the reality. This is often fragmented and piecemeal, with staff perhaps suspicious that their leaders are pursuing some deep manipulative strategy, when in reality the latter may have unwittingly allowed the process to slip out of control and in some cases been less than sensitive to employees' misgivings. In assessing both the climate and potential for resistance to change, as well as the firm's readiness to engage in transition, the following formula provides us with a basis for gaining insight.

$$C = (a\ b\ d) > x$$

where C = change; a = dissatisfaction with the status quo; b = clear or understood desired state; d = practical first steps towards the desired state; and x = the cost of changing.

Dissecting this model, it is evident that to effect change, there should exist within the firm sufficiently widespread dissatisfaction with existing systems and procedures to foster and mobilize the necessary energy for change. This has often been found lacking in some organizations, with employees visibly apathetic to change. In some instances the latter may have expressed a strong desire to maintain the status quo, confident that existing methods of working, were appropriate and efficient. Consequently they may need to be convinced that a move to quality management motivated by client/customer pressure would also yield the necessary level of internal efficiencies, to justify such upheaval. It is also important to appreciate that such transition processes will need to run in parallel with maintaining existing customer services, greatly adding to employees' workload. Hence convincing and realistic arguments must be provided by senior management to mitigate any potential resistance.

An important element of the model is the presentation of a clear image and conception of what the state of affairs will be like following successful transition. Since this is often the first time that senior managers will have effected this type of change to quality management, there may be little in their immediate frame of reference which can help. In which case, information external to the firm must be sought, from published reports, research, conferences and the experiences of others who have undergone the process. Misconceptions and false promises of overwhelming benefits should be tempered with a more realistic and sustainable vision. The results of the CIRIA research in 1995 demonstrate the apparent mismatch between expected and actual benefits perceived by employees as accruing consequent to certification.

The model also indicates that there should be clear consensus on what first steps are necessary to effect the change process. Earlier in this chapter, it was mentioned that CIRIA found a lack of clarity in the methodologies of some of the firms before implementation of the QMS. Finally, the formula indicates that the costs of changing need to be outweighed by the aforementioned three elements. Costs include not only financial and resource-based expenditure, but also their indirect impact on existing values systems and behaviours. The first, more objective measures of costs are difficult enough to gauge even after the event, the latter less tangible costs, may have both immediate and long-term consequences.

Notwithstanding the above, while such changes are being implemented, firms still need to function to produce and to offer their services. The change to quality management may also coincide with other initiatives such as downsizing, re-engineering, etc. further complicating a complex process.

20.4 CONCLUSIONS

Misgivings have been expressed about firms deciding to install a system of quality management based on BS 5750/ISO 9000, as their initial and sole response to quality management. Analysis of evidence has revealed cases where motivation to change may be governed, not primarily by the search for internal efficiency, but more by the desire to maintain market share. Referring to the standard, Binney expressed the view that:

> . . . it is a strange place for most organisations to start the implementation of quality. ISO 9000 has an important supporting role to play in quality, provided it is handled correctly. It is not, in our view, for most companies the place to begin.

Some of the larger multinational companies have responded by pursuing in-house-led TQM programmes and consequently integrating certification to the standard as part of this process, not the complete package.

Certification to BS 5750/ISO 9000 is about conformance to requirements and the installation of confidence building systems and procedures. As a 'horizontal standard' it allows an individual interpretation within a framework and promotes introspection. It internalizes quality at the expense of the customer. It does not guarantee that the goods and services provided are quality. In contrast, TQM focuses on fulfilling customer requirements and expectations. Somewhere, the distinction between the certification to the standard and approach to TQM may have become blurred. Wider expectations may have been fuelled by a process which should never have been expected to fulfil them. This perceived mismatch may have led some employees into thinking that senior management has sold them the wrong package, fostering their disillusionment and making them intransigent to further change. Consequently, any future move to enhance quality by whatever means will be gauged against the yardstick of a previous attempt which may have mitigated against competitiveness, and being perceived by employees as misapplied, and misdirected.

20.5 REFERENCES AND FURTHER READING

Batchelor, C. (1993) A Victim of its Own Success. *Financial Times*, 21 July.

Beckhard, R. Kolb, D. Rubin, *et al.* (1990) *Strategies for Large System Change in Organisational Behaviour: Practical Readings for Managers*, Prentice-Hall, Englewood Cliffs, NJ.

Binney, G. (1992) *Making Quality Work: Lessons from Europe's Leading Companies*, Special Report No.p655, The Economist Intelligence Unit.

Bleer, M. and Eisenstat, R.A. (1990) *Critical Path for Corporate Review*, Harvard Business School.

Boisot, M. (1987) *Information and Organisations*, Fontana, London.

CIRIA (1995) Research Project RP455, Quality Management in Construction, Survey of Experiences with BS 5750, CIRIA.

Kolb, D., Rubin, I. and Osland, J. (1991) *Organisational Behaviour: An Experiential Approach*, 5th edn, Prentic-Hall, Englewood Cliffs, NJ.

Lai, Hon-Keung (1989) Integrating total quality and buildability: a model for success in construction; CIOB Technical Paper No. 109.

Lansley, P.L. (1994) Analysing construction organisations. *Construction Management and Economics*, **12**, 337–48.

O'Brien, P. (1992) Low Adoption Rate of Quality Standards in the Service Sector; Proceedings, 2nd Workshop on Quality Management in Services, European Institute for Advanced Studies in Management.

RIBA (1990) *Quality Management – Guidance for an Office Manual; Section* 4c, RIBA.

RICS (1992) *Quality Management in Organisations employing Chartered Surveyors*, RICS (unpublished).

CHAPTER 21

Can quality work in construction? Its effects on contracting firms

21.1 INTRODUCTION

This chapter describes the effects of introducing quality initiatives such as ISO 9000 (formerly BS 5750) and TQM (Total Quality Management) in ten large contracting firms.

The emphasis is to analyse the effectiveness of these initiatives on the main constituent of the ten firms – people. Quality has been implemented within the context of much hype with regard to its potency. It is our hope in this chapter to provide an insight into the difficulties that practitioners face, most particularly quality managers, in the introduction and maintenance of such initiatives.

We should also pay some attention to the way in which the industry operates, in that construction firms have an inherent culture which makes improvement in procedures difficult? How can those vested with the task of bringing about change deal with the various resistances that will naturally occur? Finally, what is the future as far as these ten firms are concerned? Can they bring about any radical improvement in the same way as has occurred in other industries? In other words, can they become 'World Class'?

This is therefore an account of the issues that the introduction of quality has brought about in ten contractors using a qualitative methodology. It does so primarily from the viewpoint of those most directly involved in its implementation and management: the Quality Managers. These individuals have the task of bringing about change in what are often very difficult and trying conditions, as resistance to change is a very real hurdle in an industry which is undergoing the biggest recession and structural downsizing in its history. Resentment is generated among those who see their routines disturbed or their jobs at risk. The Quality Managers or Change Agents (CA) as they will be referred to have had to develop ways of dealing with these concerns;

they must develop 'coping strategies' in order to make the introduction of change more palatable. Ultimately, they have to demonstrate to the directors and shareholders that quality initiatives will bring long-term benefit and hence value for money to their organizations.

Each of the ten firms are at different stages in the development of their quality programmes, although all are identical in having obtained registration to ISO 9000. Some are in the process of developing TQM, a generic title which covers many varying approaches while others are planning to do so after a period of consolidation. The rest view achieving formal QA (Quality Assurance) to ISO 9000 as having been a significant hurdle in itself and are concerned at the introduction of further initiatives of which cultural change is a prerequisite.

21.2 QUALITY AND ITS CONTEXT IN THE UK CONSTRUCTION INDUSTRY

The requirement for QA was initially brought about by pressure from large clients, particularly in the public sector. It is recognized that the influence of the PSA (Property Services Agency of the Department of the Environment) was a prime mover behind many large contractors becoming registered to BS 5750, as there was a fear that failure to do so would result in removal from tender lists. This, coupled with the provision of grants by the government to assist in QA consultancy advice, caused a rush for registration in the late 1980s.

While the perceived need to register can be viewed in retrospect as having been done under duress, the potential benefits seemed significant. These appeared in a plethora of publications which extolled the virtues of Quality Assurance. Typical were Hughes and Williams, who argued that it was required for commercial, marketing and organizational reasons and major advantages would be manifested through improved communications and efficiency. Mistakes would be avoided and projects would be more likely to be completed on time. The message was clear that QA would save your organization money and help to make increased profit. The belief seems to have been that once all contractors operated Quality Systems there would be widespread improvement in the industry. Further 'trickle down' was expected whereby all contractors would require their sub-contractors and suppliers to be operating to a formal quality system.

The need to be part of the quality movement seems to be essential to contemporary business. The allure was too great to resist: sustainability; achievement of customer satisfaction; enhanced market image; and, possibly most important of all, more profit. Not surprisingly construction became part of the trend. No organization could sensibly afford not to; it was too good to miss.

21.3 QA: THE STORY SO FAR – GOOD AND BAD

The work carried out so far has indicated that in each of the ten firms considered, the main drive was to get registered. This was described by one CA as the 'Plaque on the wall syndrome'. What was meant was that in the rush to get registered, people tended to forget that what was being implemented was an administrative system – as such it was for many of the firms a refinement of existing procedures. For some it was the first time that a definitive manual of operations had been actually produced. The danger that appeared to concern many of those who were responsible for implementing the systems was that QA would produce 'instant jam'.

On balance, all of those questioned favour some form of formal quality assurance system. Its introduction did not produce the benefits that some of the advocates had predicted, but the mere act of putting in a formal system had highlighted areas of concern previously noticed by the firm. Typical is the following account:

> Before its implementation each year we would knock out a couple of thousand cubic metres of concrete. This was because it was in the wrong position, incorrect in dimension or not properly specified. I knew this was something that could be improved by QA. So I deliberately wrote procedures which sought to make people think more carefully about what they were doing. There were hold points to get things double checked. I got a lot of criticism from those who said that this would cause problems by stopping the concrete gang. But I told them 'If we continue to make the same mistakes we'll go out of business'. Besides I asked 'Do you have any better ideas?' Five years later we now knock out a couple of dozen metres. That's a huge saving in labour and materials. It has justified my belief and added tremendously to our ability to survive. I tell people that this is what quality is really about – adding value to our processes. The consequence is that I have greater credibility and people are much more willing to listen and better still contribute ideas for improvement.

This certainly indicated that the use of a formal system of work had produced benefit to this particular company. What seemed to be the most important was the approach of those vested with the task. All bemoan their inability to demonstrate hard measures beyond indications. Quality costing as a concept advocated by Dale and Plunkett has yet to be adopted as a widespread practice which often leaves the CAs in the difficult position of having to defend continued registration. Many indicate their dissatisfaction with those clients who demand that only firms who have QA can tender. But often they do not implement the threat properly or fail to audit them. This leads some to feel the whole efforts has been a sham:

> I don't believe that I'm untypical when I say that we (as an industry) were led up the garden path. There were some who said that clients would only use QA firms. As a result we could charge a bit more to cover the extra cost. But the reality is that they want you to have it before they'll consider your tender and then give the work to someone who is far cheaper. Often these firms have a Mickey Mouse system which is enough to satisfy the client. They also tend to be the sort who will use anybody as long as they are cheap. I am now having to defend our continued involvement in QA. I wish we had some measures which could prove that some good has come of it.

One very positive effect was identified in all those companies studied. There is an increased confidence among the CAs that having developed a systematic method of working positive change is now possible. The firms seem to be able to take a less timid stand when dealing with clients or their advisors. As one CA stated:

> In the past we took a lot of unjustified criticism from some clients and architects. The fact that they were talking nonsense didn't matter, we let them walk all over us. They seemed to think we were uneducated builders who they could treat as they pleased. To be honest we also didn't have much esteem in our own abilities. This led to a pretty unhappy state of affairs. When a problem occurred which could be to our advantage we'd have no hesitation about hitting them with claims. Now we have a quality system we are much more well equipped to say no to certain things. It's all about respect. If we are treated fairly we will be more willing to let something go. Let's face it, claims often cause more problems than they're worth.

Another was positively glowing about what had been achieved:

> We have certainly been able to impress some big clients with both the quality of the finished product and the way we present ourselves. Our documentation is really professional. This is something that was not regarded as being important. One high street chain has said they were really impressed with us and would have no hesitation in recommending us. This is down to QA. It has forced us to confront all the woolly and shady practices that we accepted as being the 'way of the industry'. It led to us making some incredible mistakes in the past. This was no longer acceptable and we had to get our act together.

As might be expected with any change initiative there have been problems. One of the complaints most commonly heard is that QA has increased bureaucracy because of the requirement of third-party auditing which necessitates procedures records and systematic methods of carrying out functions. The result has led to comments such as the following made by a site manager:

> I agree with the principles of QA. But all we see on site are more and more forms to fill in and procedures to be adhered to. Before it was implemented I spent 99% of my time out there. I made sure everything was done properly. Now I spend at 10–15% in here doing the paperwork for QA. If something is done wrong or not to my liking it takes longer and is more costly to put right. I've got all the paperwork to say everything is right, but at what cost?

Many also viewed QA as a necessary evil, in that they saw that its imposition was essential to avoid being excluded from tender lists. But the need to have third-party auditors was regarded as a distraction at best, and at worst a 'pantomime' as one called it or a downright interference which causes harm. One site manager was very certain in his views:

> You get them (third party auditors) coming here for the day. They know nothing at all about construction. They ask stupid questions and want to see meaningless pieces of paper. They are usually the pieces of paper that we don't use on a day to day basis and so are impossible to find. It is very time consuming. And when its all over you get a list of non conformances which the contracts manager uses against you.

For many of those interviewed there was a recognition that QA has caused concern and disruption and that further change needs to be handled delicately. It is accepted that future initiatives have to be based upon the expertise of those at the heart of any organization – the people. The following comment reflects this belief:

> The people out there are sick of the jargon. They've had enough of QA. What they see as being important is to their job well. They will go along with it as long as they can see benefit. What we, as a professional outfit, have to be about is being seen to be better than the rest. In some ways we need to go back to the old days. We need to regain the pride we had in doing a good job and pleasing the client. Probably a major part will be to stop using language which alienates and threatens our best asset – our employees.

21.4 QUALITY – THE ROLE OF THE CHANGE AGENT

Many of those taking part in this study spoke of their ability to deal with the concerns of people in their organizations. Many spoke of the need to develop trust and cooperation, as without this they felt that any improvement initiative would not have any foundation and be destined to collapse. They all agree on the necessity of having tacit support at every level of the organization and all decry the use of threats as a way to introduce change and gain acceptance. The following account demonstrates the need for sensitivity:

> Sure you can wave the big stick and say do it or I'll report you to senior management. But when we first started QA the Managing Director made a statement that it was required at all costs and that any resistance would not be tolerated. I thought 'That will really get the troops rallying to the cause, I don't think!' It took a lot of effort to lessen this perception. I'm sure there are some who still think I'm a management lackey. But you have to be really careful and treat everyone as you find them. The old timers are the worst. You have to translate everything into traditional language because many of them say 'QA is nothing new. I've been doing it all my life'.

The task of preparing the actual procedures gives an opportunity to get people to become involved in the process of improvement. Forcing people to work to procedures not written by themselves is recognized as a recipe for certain failure in any service organization which relies on the skill and judgement of people producing one-off items, like construction projects:

> You have to get everyone to buy into the system. This is often not very easy because they say 'That's your job'. You tell us what to do and we'll follow the procedures. But the probability is that they'll do anything to avoid using them as a tool for improvement. I want people to feel that it is their system. They will then do the basic element not because it is a chore but because they see value. Even better, you start to get them saying 'Why don't we try this or perhaps if we changed that . . .'. When this happens I feel that we are really starting to get somewhere.

Another commented:

> The system and procedures must reflect what really goes on. This is not as straightforward as it might appear. I've worked on site and I know that there some dodgy ways of doing things. Fair enough it is supposed to be my job to eradicate them, but how? I can't do it by arm twisting. That would completely turn them off and confirm the suspicions of some that I had sold out when I took this job. No, it's a question of getting them to think that there may be other different ways. I never say 'Do it this way because I think it is better'. They have to make that decision for themselves. In

> most cases they eventually will change. My philosophy is that our QA system has to reflect the diversity of approaches of our employees.

This attitude is typical of the approach of those change agents, who contributed. It appears to be a form of 'give and take' as there is an acceptance that change comes slowly, contrary to the belief of senior management. They know that a large part of what they do is concerned with human relationships and not merely plugging in a completely automated technical system.

21.5 FROM QA TO TQM – LEARNING HOW TO DEAL WITH CULTURAL CHANGE

For many of the firms there is a desire to continue the momentum that quality assurance has generated, which is a reflection of the external influence of some clients but also a desire to capitalize on the skills and expertise of their own people. Brown suggests that gaining ISO 9000 is not the culmination of quality activities. It is a '. . . a short-sighted view which is likely to lead to disappointment and will not bring out the full benefits of a QMS'. He further believes that '. . . ISO 9000 should be viewed merely as the first stage on the never ending road to quality. This route to continuous improvement is the process commonly called TQM'.

This change is not an easy task for any organization and presents CAs with a difficult challenge. As Brown puts it:

> Essentially TQM is the philosophy behind managing a change process in such a way that there are lasting benefits. The nature of change is that it is a continuous process. Change is also something many firms and individuals at worst dislike, or at best are wary of.

Those interviewed agreed. As one said:

> Getting ISO 9000 is like riding a bike up a small hill. Once you get to the top there is a tendency to slacken off and go down hill. But you've got to keep everyone working consistently, like you're on a plateau. But moving to TQM is like coming to a mountain. You know it will be hard and will hurt.

Brown sums up perhaps the biggest problem of moving from QA to TQM when he says that trying to explain TQM is difficult, as its methodology and end results are not as clear-cut as ISO 9000, whereby a certificate awaits those who successfully stay the relatively short course.

In construction, the transition from writing procedures to changing the culture is bound to be difficult. It requires skills which are different to those employed in the pursuit of registration. This is because 'Total

Quality is a process of change and no single change is more important than that of employees' attitudes. A shift in management style is necessary to allow change to happen'.

The need for senior management commitment is the most frequently cited axiom in this process. But the reality is often that their involvement causes problems for the CAs. Their biggest concern is that there is the acceptance of the need for change but they cannot agree among themselves as to what should be done, and how; or they think it is all right to make grand statements about unity of purpose then go back to their offices and refuse to change themselves. This is also recognized as one of the major reasons why people further down the firm find change difficult to accept.

The current market in construction, as with many other industries, is one in which cost is the main determinant to survival. The conditions therefore accord with what Rampey and Roberts believe are necessary for TQM whereby organizations must put in place a

> . . . people focused management system that aims at continual increase in customer satisfaction at continually lower real cost. TQ is a total systems approach (not separate area or program), and an integral part of a high level strategy. It works horizontally across functions and departments, involving all employees top to bottom, and extends backwards and forwards to include the supply chain and the customer chain . . .

If there are problems within the individual firms in terms of developing the unity of purpose and dedication to customer needs as an essential prerequisite to TQM then achieving cooperation with sub-contractors and suppliers would seem to have even further to go. One comment summed up the dilemma:

> There is a lot of talk of partnerships between contractors and those who supply to us. But there is a deep seated mistrust. We feel that if you open up too much then you'll get stabbed in the back. I know that the TQ philosophy is based on harmony and trust but we've spent too long in 'cut throat' competition using adversarial contracts to try to win at all costs.

The influence of clients is an equally pessimistic situation. There is a hope that they will be more willing to engage constructively in the process, and by so doing they may begin realistically to understand what can be achieved. But so far the portents have not been good. Examples of clients saying one thing and doing another are distressingly abundant. As one CA related:

> We were doing a job for a large financial institution. When we got the contract they said 'Great you're working towards TQM, that

> should be beneficial to both of us'. We thought that this meant that they would play fair and square with us. But as soon as we started they started making changes and said 'We don't expect to be charged, we're the customer'. Well, there is only so much goodwill and given that they had already got us to reduce our original tender figure we were making virtually no profit. In this market that is accepted. But it is annoying when clients appear to show a commitment to TQ but still play fast and loose. They expect to make profit but don't think we need to.

Another pointed to the general feeling in the industry:

> We've had years of decline. No wage rises, no training, threats, redundancies and clients who did not play fair. We did it in turn to our sub-contractors and suppliers. Everyone is completely fed up. You can't blame them. So to talk of introducing cultural change which will make things better is met with scepticism. Most want to keep their heads down and do what they are told.

21.6 CONCLUSION: THE ARRIVAL OF A NEW PARADIGM

It is beyond argument that the construction industry is changing as clients' expectations are increasingly seen to be the only criteria by which success is measured. Organizations should now look to move to the new paradigm of providing a superior customer value strategy.

The research reported here demonstrates the need for a massive shift in the way construction operates; there has to be a move towards the key players seeing one another not in the traditional adversarial attitude but as partners to a common goal. If the laudable recommendations of the Latham report are to be implemented then more needs to be done to achieve improved communications between all participants in a construction project which would mean that greater trust and harmony would invariably result from initiatives such as partnering.

This view was reinforced by one who said:

> I've worked in other industries and seen that change is possible. It is not easy but my word the results can be spectacular. The biggest change comes about in the people. They gain confidence in their ability. This in turn leads to provision of a higher degree of professionalism in their dealings with the client downwards. John Egan at BAA has helped to bring this message. We need more like him.

Another was optimistic about the future:

> The use of the NEC (New Engineering Contract) will help to improve things. But what is necessary is that those who have

influence learn to trust each other. I mean what have they got to lose? Look at what their behaviour over the last twenty years has done to the industry.

According to work carried out by the UMIST Quality Management Centre, companies in Europe have a much more varied approach to quality. As such they identify six different levels of TQM adoption: (1) Uncommitted; (2) Drifters; (3) Tool-Pushers; (4) Improvers; (5) Award Winners; and (6) World Class. The ten companies in our survey are at various stages in their quality programmes and therefore are at different levels on this scale. Notably four are working towards the European Model for Total Quality Management, which would put them at level (5). Some are also working towards the Investors in People standards, as outlined in a previous chapter, and indicates a level of commitment towards vast improvement. It also underlines their increasing confidence in the ability to satisfy customer expectation and the potential of their employees. Much effort will need to be put into ensuring that the organizations can satisfy the criteria. But in so doing they will have reached a point in their TQM maturity where the kind of culture, values, trust, capabilities, relationship and employee involvement in their business required to win such an award have been developed – a point at which quality improvement has become Total in nature.

The next stage would then be to become 'World Class'. This is where the integration of quality improvement in the business aims to delight the customer. Williams and Bersch claimed that fewer than ten such companies existed in the world in 1989, and all were Japanese. In 1995, there are now over 500, a handful of which are entirely British. Clearly, British construction firms such as those contributing to this study have a long way to go, but believe it to be possible to become World Class.

If more justification were required Oakland, Zairi and Letza carried out a study in 1994 into companies who have implemented TQM programmes. The results showed that the majority had achieved greater success against seven performance indicators than was the average for their respective industries.

Those contractors involved in this study are sufficiently aware of what is required from them in order to contribute to the process of improvement. One thing is abundantly clear, they are tired of the relentless drive to stay in business at all costs. This they know is a road that provides no long-term value to their members. In short, they have to become World Class – to get better or get beaten.

21.7 REFERENCES AND FURTHER READING

BBC (1993) *BS 5750/Kaizen – Get Better or Get Beaten*, shown on BBC2 television, August.

Bell, D., McBride, P. and Wilson, G. (1994) *Managing Quality*, Butterworth-Heinemann.
Bounds, G., Yorks, L., Adams, M. and Rainey, G. (1994) *Beyond Total Quality Management – Towards the Emerging Paradigm*, McGraw-Hill.
Brown, T. (1993) *Understanding BS 5750 and other Quality Systems*, Gower, Aldershot.
Dale, B.G. and Plunkett, J.J. (1991) *Quality Costing*, Chapman & Hall, London.
Dale, B.G., Lascelles, D.M. and Boaden, R.J. (1994) Levels of total quality management adoption, in *Managing Quality* (ed. B.G. Dale), Prentice-Hall, New Jersey.
Hellard, R. (1995) *Project Partnering – Principles and Practice*, Thomas Telford. London.
Hughes, T. and Williams, T. (1991) *Quality Assurance – a Framework to Build On*, BSP Professional.
Kreikemeier, K.G. (1993) Quality – a Bottom Line Issue for Top Management, in *Proceedings of the 1st International Architectural/Engineering & Construction Division Conference*, October, San Diego, California, USA.
Latham, M. (1994) *Constructing the Team*, HMSO, London.
McCabe, S., Rooke, J. and Seymour, D. (1995) Quality Managers and Cultural Change, in *Proceedings of the 11th Annual Conference of ARCOM*, University of York, September.
Oakland, J.S., Zairi, M. and Letza, S. (1994) TQM and Bottom Line Results. *Quality World*, September.
Rampey, J. and Roberts, H. (1992) Perspectives on Total Quality, in *Proceedings of Total Quality Forum IV*, November, Cincinnati, Ohio, USA.
Williams, R. and Bersch, B. (1989) *Proceedings of the First European Quality Management Forum*, European Foundation for Quality Management, pp. 163–72.

Index